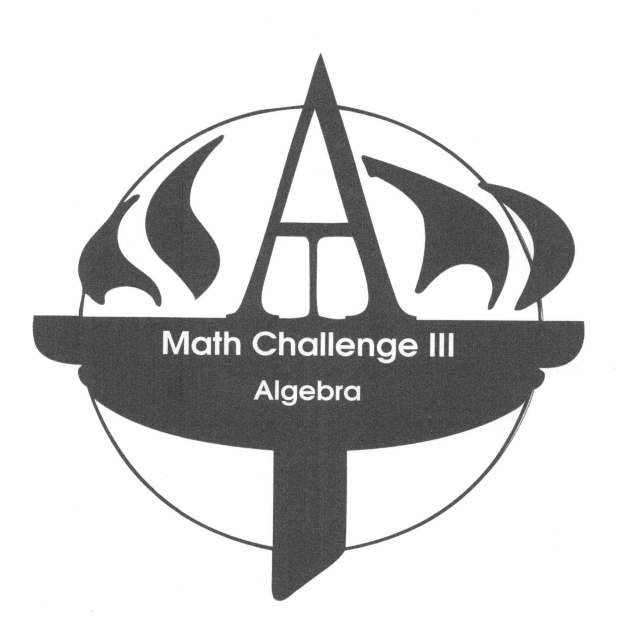

Math Challenge III
Algebra

Areteem Institute

Math Challenge III Algebra

Edited by   Kevin Wang
              David Reynoso
              John Lensmire
              Kelly Ren

Copyright © 2019 ARETEEM INSTITUTE

WWW.ARETEEM.ORG

PUBLISHED BY ARETEEM PRESS

ISBN: 1-944863-26-5
ISBN-13: 978-1-944863-26-5
First printing, March 2019.

TITLES PUBLISHED BY ARETEEM PRESS

Cracking the High School Math Competitions (and Solutions Manual) - Covering AMC 10 & 12, ARML, and ZIML
Mathematical Wisdom in Everyday Life (and Solutions Manual) - From Common Core to Math Competitions
Geometry Problem Solving for Middle School (and Solutions Manual) - From Common Core to Math Competitions
Fun Math Problem Solving For Elementary School (and Solutions Manual)

ZIML MATH COMPETITION BOOK SERIES

ZIML Math Competition Book Division E 2016-2017
ZIML Math Competition Book Division M 2016-2017
ZIML Math Competition Book Division H 2016-2017
ZIML Math Competition Book Jr Varsity 2016-2017
ZIML Math Competition Book Varsity Division 2016-2017
ZIML Math Competition Book Division E 2017-2018
ZIML Math Competition Book Division M 2017-2018
ZIML Math Competition Book Division H 2017-2018
ZIML Math Competition Book Jr Varsity 2017-2018
ZIML Math Competition Book Varsity Division 2017-2018

MATH CHALLENGE CURRICULUM TEXTBOOKS SERIES

Math Challenge I-A Pre-Algebra and Word Problems
Math Challenge I-B Pre-Algebra and Word Problems
Math Challenge I-C Algebra
Math Challenge II-A Algebra
Math Challenge II-B Algebra
Math Challenge III Algebra
Math Challenge I-A Geometry
Math Challenge I-B Geometry
Math Challenge I-C Topics in Algebra
Math Challenge II-A Geometry
Math Challenge II-B Geometry
Math Challenge III Geometry
Math Challenge I-A Counting and Probability
Math Challenge I-B Counting and Probability
Math Challenge I-C Geometry

Math Challenge II-A Combinatorics
Math Challenge II-B Combinatorics
Math Challenge III Combinatorics
Math Challenge I-A Number Theory
Math Challenge I-B Number Theory
Math Challenge I-C Finite Math
Math Challenge II-A Number Theory
Math Challenge II-B Number Theory
Math Challenge III Number Theory

COMING SOON FROM ARETEEM PRESS

Fun Math Problem Solving For Elementary School Vol. 2 (and Solutions Manual)
Counting & Probability for Middle School (and Solutions Manual) - From Common Core to Math Competitions
Number Theory Problem Solving for Middle School (and Solutions Manual) - From Common Core to Math Competitions

The books are available in paperback and eBook formats (including Kindle and other formats).
To order the books, visit https://areteem.org/bookstore.

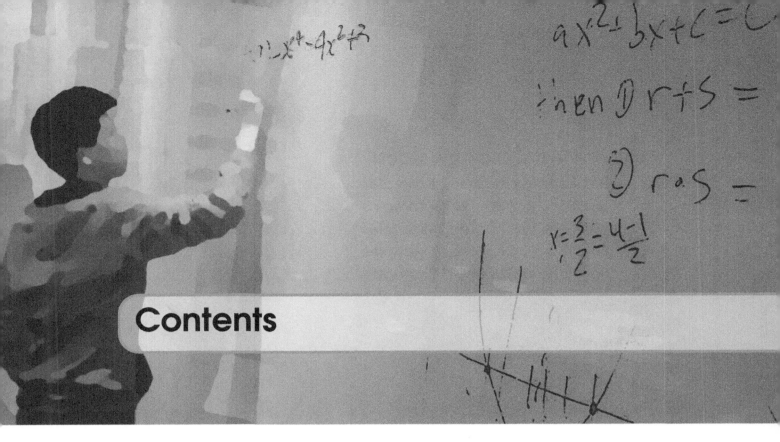

Contents

Introduction 7

1   Review of Logarithm 15
1.1 Example Questions 17
1.2 Practice Questions 20

2   Fundamentals of Complex Numbers 25
2.1 Example Questions 30
2.2 Practice Questions 33

3   Solving Equations 37
3.1 Example Questions 40
3.2 Practice Questions 43

4   Solving Inequalities 47
4.1 Example Questions 51
4.2 Practice Questions 55

5   Special Algebraic Techniques 59
5.1 Example Questions 64
5.2 Practice Questions 68

Solutions to the Example Questions 73

1   Solutions to Chapter 1 Examples 74
2   Solutions to Chapter 2 Examples 82
3   Solutions to Chapter 3 Examples 91
4   Solutions to Chapter 4 Examples 98
5   Solutions to Chapter 5 Examples 110

Introduction

The math challenge curriculum textbook series is designed to help students learn the fundamental mathematical concepts and practice their in-depth problem solving skills with selected exercise problems. Ideally, these textbooks are used together with Areteem Institute's corresponding courses, either taken as live classes or as self-paced classes. According to the experience levels of the students in mathematics, the following courses are offered:

- Fun Math Problem Solving for Elementary School (grades 3-5)
- Algebra Readiness (grade 5; preparing for middle school)
- Math Challenge I-A Series (grades 6-8; intro to problem solving)
- Math Challenge I-B Series (grades 6-8; intro to math contests e.g. AMC 8, ZIML Div M)
- Math Challenge I-C Series (grades 6-8; topics bridging middle and high schools)
- Math Challenge II-A Series (grades 9+ or younger students preparing for AMC 10)
- Math Challenge II-B Series (grades 9+ or younger students preparing for AMC 12)
- Math Challenge III Series (preparing for AIME, ZIML Varsity, or equivalent contests)
- Math Challenge IV Series (Math Olympiad level problem solving)

These courses are designed and developed by educational experts and industry professionals to bring real world applications into the STEM education. These programs are ideal for students who wish to win in Math Competitions (AMC, AIME, USAMO, IMO,

ARML, MathCounts, Math League, Math Olympiad, ZIML, etc.), Science Fairs (County Science Fairs, State Science Fairs, national programs like Intel Science and Engineering Fair, etc.) and Science Olympiad, or purely want to enrich their academic lives by taking more challenges and developing outstanding analytical, logical thinking and creative problem solving skills.

The Math Challenge III (MC III) courses are for students who are qualified to participate in the AIME contest, or at the equivalent level of experience. The MC III topics include polynomials, inequalities, special algebraic techniques, triangles and polygons, coordinates, numbers and divisibility, modular arithmetic, advanced counting strategies, binomial coefficients, sequence and series, complex numbers, trigonometry, logarithms, and various other topics, and the focus is more on in-depth problem solving strategies, including pairing, change of variables, advanced techniques in number theory and combinatorics, advanced probability theory and techniques, geometric transformations, etc. The curricula have been proven to help students develop strong problem solving skills that make them perform well in math contests such as AIME, ZIML, and ARML.

The course is divided into four terms:

- Summer, covering Algebra
- Fall, covering Geometry
- Winter, covering Combinatorics
- Spring, covering Number Theory

The book contains course materials for Math Challenge III: Algebra.

We recommend that students take all four terms, but the terms do not build on previous terms, so they do not need to be taken in order and students can take single terms if they want to focus on specific topics.

Students can sign up for the course at `classes.areteem.org` for the live online version or at `edurila.com` for the self-paced version.

About Areteem Institute

Areteem Institute is an educational institution that develops and provides in-depth and advanced math and science programs for K-12 (Elementary School, Middle School, and High School) students and teachers. Areteem programs are accredited supplementary programs by the Western Association of Schools and Colleges (WASC). Students may attend the Areteem Institute in one or more of the following options:

- Live and real-time face-to-face online classes with audio, video, interactive online whiteboard, and text chatting capabilities;
- Self-paced classes by watching the recordings of the live classes;
- Short video courses for trending math, science, technology, engineering, English, and social studies topics;
- Summer Intensive Camps held on prestigious university campuses and Winter Boot Camps;
- Practice with selected free daily problems and monthly ZIML competitions at ziml.areteem.org.

Areteem courses are designed and developed by educational experts and industry professionals to bring real world applications into STEM education. The programs are ideal for students who wish to build their mathematical strength in order to excel academically and eventually win in Math Competitions (AMC, AIME, USAMO, IMO, ARML, MathCounts, Math Olympiad, ZIML, and other math leagues and tournaments, etc.), Science Fairs (County Science Fairs, State Science Fairs, national programs like Intel Science and Engineering Fair, etc.) and Science Olympiads, or for students who purely want to enrich their academic lives by taking more challenging courses and developing outstanding analytical, logical, and creative problem solving skills.

Since 2004 Areteem Institute has been teaching with methodology that is highly promoted by the new Common Core State Standards: stressing the conceptual level understanding of the math concepts, problem solving techniques, and solving problems with real world applications. With the guidance from experienced and passionate professors, students are motivated to explore concepts deeper by identifying an interesting problem, researching it, analyzing it, and using a critical thinking approach to come up with multiple solutions.

Thousands of math students who have been trained at Areteem have achieved top honors and earned top awards in major national and international math competitions, including Gold Medalists in the International Math Olympiad (IMO), top winners and qualifiers at the USA Math Olympiad (USAMO/JMO) and AIME, top winners at the

Zoom International Math League (ZIML), and top winners at the MathCounts National Competition. Many Areteem Alumni have graduated from high school and gone on to enter their dream colleges such as MIT, Cal Tech, Harvard, Stanford, Yale, Princeton, U Penn, Harvey Mudd College, UC Berkeley, or UCLA. Those who have graduated from colleges are now playing important roles in their fields of endeavor.

Further information about Areteem Institute, as well as updates and errata of this book, can be found online at http://www.areteem.org.

About Zoom International Math League

The Zoom International Math League (ZIML) has a simple goal: provide a platform for students to build and share their passion for math and other STEM fields with students from around the globe. Started in 2008 as the Southern California Mathematical Olympiad, ZIML has a rich history of past participants who have advanced to top tier colleges and prestigious math competitions, including American Math Competitions, MATHCOUNTS, and the International Math Olympaid.

The ZIML Core Online Programs, most available with a free account at ziml.areteem.org, include:

- **Daily Magic Spells:** Provides a problem a day (Monday through Friday) for students to practice, with full solutions available the next day.
- **Weekly Brain Potions:** Provides one problem per week posted in the online discussion forum at ziml.areteem.org. Usually the problem does not have a simple answer, and students can join the discussion to share their thoughts regarding the scenarios described in the problem, explore the math concepts behind the problem, give solutions, and also ask further questions.
- **Monthly Contests:** The ZIML Monthly Contests are held the first weekend of each month during the school year (October through June). Students can compete in one of 5 divisions to test their knowledge and determine their strengths and weaknesses, with winners announced after the competition.
- **Math Competition Practice:** The Practice page contains sample ZIML contests and an archive of AMC-series tests for online practice. The practices simulate the real contest environment with time-limits of the contests automatically controlled by the server.
- **Online Discussion Forum:** The Online Discussion Forum is open for any comments and questions. Other discussions, such as hard Daily Magic Spells or the Weekly Brain Potions are also posted here.

These programs encourage students to participate consistently, so they can track their progress and improvement each year.

In addition to the online programs, ZIML also hosts onsite Local Tournaments and Workshops in various locations in the United States. Each summer, there are onsite ZIML Competitions at held at Areteem Summer Programs, including the National ZIML Convention, which is a two day convention with one day of workshops and one day of competition.

ZIML Monthly Contests are organized into five divisions ranging from upper elementary school to advanced material based on high school math.

- **Varsity:** This is the top division. It covers material on the level of the last 10 questions on the AMC 12 and AIME level. This division is open to all age levels.
- **Junior Varsity:** This is the second highest competition division. It covers material at the AMC 10/12 level and State and National MathCounts level. This division is open to all age levels.
- **Division H:** This division focuses on material from a standard high school curriculum. It covers topics up to and including pre-calculus. This division will serve as excellent practice for students preparing for the math portions of the SAT or ACT. This division is open to all age levels.
- **Division M:** This division focuses on problem solving using math concepts from a standard middle school math curriculum. It covers material at the level of AMC 8 and School or Chapter MathCounts. This division is open to all students who have not started grade 9.
- **Division E:** This division focuses on advanced problem solving with mathematical concepts from upper elementary school. It covers material at a level comparable to MOEMS Division E. This division is open to all students who have not started grade 6.

The ZIML site features are also provided on the ZIML Mobile App, which is available for download from Apple's App Store and Google Play Store.

Acknowledgments

This book contains many years of collaborative work by the staff of Areteem Institute. This book could not have existed without their efforts. Huge thanks go to the Areteem staff for their contributions!

The examples and problems in this book were either created by the Areteem staff or adapted from various sources, including other books and online resources. The original resources are credited whenever possible. However, it is not practical to list all such resources. We extend our gratitude to the original authors of all these resources.

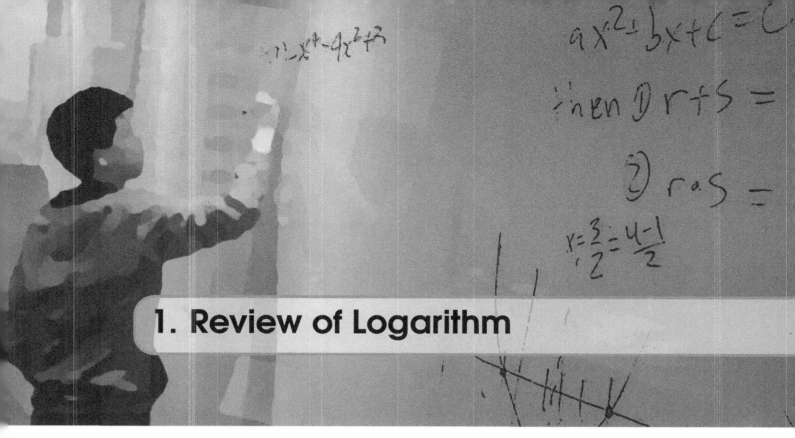

1. Review of Logarithm

Fundamentals of Logarithm

Logarithms are the inverse of exponentials. For example, given the base 2 and a number 1024, we want to find the value b so that $2^b = 1024$. Since we know that $2^{10} = 1024$, the answer is $b = 10$. This process can be written in logarithm as

$$\log_2 1024 = 10.$$

In general, if $a^b = x$, then $\log_a x = b$.

- Domain of definition: $a > 0$ and $a \neq 1$; $x > 0$
- Special values: $\log_a 1 = 0$; $\log_a a = 1$.
- Identities: $a^{\log_a x} = x$; $\log_a a^b = b$.
- Operational properties:
  - (i) $\log_a xy = \log_a x + \log_a y$
  - (ii) $\log_a \dfrac{x}{y} = \log_a x - \log_a y$
  - (iii) $\log_a x^n = n \log_a x$
  - (iv) $\log_a \sqrt[n]{x} = \dfrac{1}{n} \log_a x$

Logarithm and Exponentiation

Logarithm	Exponentiation
$\log_a x = b$	$a^b = x$
$\log_a xy = \log_a x + \log_a y$	$a^b \cdot a^c = a^{b+c}$
$\log_a x^n = n \log_a x$	$(a^b)^n = a^{bn}$
$\log_a 1 = 0$	$a^0 = 1$
$\log_a a = 1$	$a^1 = a$
$\log_a a^b = b$	$a^{\log_a x} = x$

Change of Base

Let $a > 0, a \neq 1$, $b > 0$, $b \neq 1$, and $x > 0$, then

$$\log_a x = \frac{\log_b x}{\log_b a}.$$

In the formula above, if we let $x = b$, then

$$\log_a b = \frac{1}{\log_b a}.$$

More properties:

(i) $\log_{a^n} x^m = \dfrac{m \log_b x}{n \log_b a}$

(ii) $\log_{\sqrt[n]{a}} \sqrt[m]{x} = \dfrac{n \log_b x}{m \log_b a}$

(iii) $\log_{a^n} x^n = \log_a x$

Particular Bases

(i) Common logarithm: base 10, often written as $\lg x$. In other words, $\lg x = \log_{10} x$.

    (a) If N is a positive integer, $\lfloor \lg N \rfloor$ equals the number of N's digit minus 1. Here $\lfloor x \rfloor$ means the greatest integer not exceeding x.

    Proof. Let k be the number of digits of N, then $10^{k-1} \le N < 10^k$, therefore $k - 1 \le \lg N < k$, which means $\lfloor \lg N \rfloor = k - 1$. ∎

    (b) If $0 < x < 1$, expressed in base 10, and the first non-zero digit occurs at the kth digit after the decimal point, then $\lfloor \lg x \rfloor = -k$. For example, $\lfloor \lg 0.00261 \rfloor = -3$.

(ii) Natural logarithm: base e, where e is Euler's number: $e = 2.718281828459045\ldots$. Often written as $\ln x$. In other words, $\ln x = \log_e x$.
Natural logarithm is widely used in mathematics, physics, engineering, etc.

(iii) Binary logarithm: base 2.
Binary logarithm is commonly used in computer science and engineering.

Logarithm Functions

Let $f(x) = \log_a x$ be a function where $a > 0$ and $a \neq 1$.

(i) The function $f(x)$ has domain $(0, +\infty)$ and range $(-\infty, +\infty)$.
(ii) $f(x)$ is an increasing function if $a > 1$, a decreasing function if $0 < a < 1$.
(iii) The graph of $y = f(x)$ has a vertical asymptote $x = 0$.
(iv) The graph of $y = f(x)$ intersects the x-axis at $(1, 0)$.

1.1 Example Questions

Problem 1.1 Show that

$$x^{\log_a y} = y^{\log_a x}$$

Remark

The identity in Problem 1.1 is a very nice one. We will use it later on in other examples.

Problem 1.2 Assume function $f(x) = \log_2(x^2 + ax + 1)$ is well-defined on all $x \in \mathbb{R}$. Find the range of possible values of a.

Problem 1.3 Let $x > 0$ be a real number. Given that $\log_{\sqrt{2}} x = 100$, compute $\log_{\sqrt{x}} 2$.

Problem 1.4 Solve the equation for x:

$$(\log_5 x)^2 = \log_5 x^2.$$

Problem 1.5 Let $a > 0$ and $a \neq 1$, Show that $\log_{a^2} X = \dfrac{3}{2} \log_{a^3} X$ for all $X > 0$.

Problem 1.6 Show that $\log_{10} 2$ is irrational.

Problem 1.7 Evaluate

$$\frac{5}{\log_2 2016^3} + \frac{2}{\log_3 2016^3} + \frac{1}{\log_7 2016^3}.$$

Problem 1.8 Given that $1 < a < b < a^2$, arrange the following four numbers in increasing order:

$$2, \quad \log_a b, \quad \log_b a, \quad \log_{ab} a^2.$$

Problem 1.9 Evaluate $\lg\left(\sqrt{3+\sqrt{5}}+\sqrt{3-\sqrt{5}}\right)$. Here lg means logarithm with base 10.

Problem 1.10 Evaluate $6^{\lg 40} \times 5^{\lg 36}$. Here lg represents logarithm with base 10.

Problem 1.11 The real numbers x, y, and z are all greater than 1, and w is a positive number such that $\log_x w = 45$, $\log_y w = 60$, and $\log_{xyz} w = 15$. Find $\log_z w$.

Problem 1.12 Let x_1 be a root of the equation $\log_3 x + x - 3 = 0$, and x_2 be a root of the equation $3^x + x - 3 = 0$, find the value of $x_1 + x_2$.

Problem 1.13 Given that a, b, c are all positive and not equal to 1, simplify the following:

$$a^{\log_2(b/c)} \cdot b^{\log_2(c/a)} \cdot c^{\log_2(a/b)}$$

Problem 1.14 Let x and y be positive and $\log_9 x = \log_{12} y = \log_{16}(x+y)$, find $\dfrac{y}{x}$.

Problem 1.15 Let m and n be positive integers, $a > 0$ and $a \neq 1$, and

$$\log_a m + \log_a\left(1 + \frac{1}{m}\right) + \log_a\left(1 + \frac{1}{m+1}\right) + \cdots + \log_a\left(1 + \frac{1}{m+n-1}\right)$$
$$= \log_a m + \log_a n,$$

find the values of m and n.

1.2 Practice Questions

Problem 1.16 Given that a, b, c are all positive numbers and not equal to 1, simplify the following:

$$\log_a b \cdot \log_b c \cdot \log_c a$$

Problem 1.17 Given that $\log_4 x = 15$, find the value of $\log_8 x$.

Problem 1.18 If $\log_x 5 = 3$, what is x^9?

Problem 1.19 Given that $\log_{10} 2 \approx 0.3010$. Expressed in base 10, how many digits does the number 2^{100} have?

Problem 1.20 Let $a = \log_7(5\sqrt{2} - 1) + \log_2(\sqrt{2} + 1)$, and $b = \log_7(5\sqrt{2} + 1) + \log_2(\sqrt{2} - 1)$, find the value of $a + b$.

Problem 1.21 Evaluate $\log_2\left(\sqrt{3 + \sqrt{5}} - \sqrt{3 - \sqrt{5}}\right)$.

Problem 1.22 Solve the equation for x:

$$8^{\log_6(x^2 - 7x + 15)} = 5^{\log_6 8}.$$

Problem 1.23 Solve the equation for x: $\log_{10}\log_{100}x = \log_{100}\log_{10}x$.

Problem 1.24 Given that $\log_{10}3 \approx 0.477121$. If we convert the base-10 number 10^{100} into base 3, how many digits does it have?

Problem 1.25 Assume $x > 2, y > 2$, and let $a = \lfloor\log_2 x\rfloor, b = \{\log_2 x\}, c = \lfloor\log_2 y\rfloor, d = \{\log_2 y\}$. Also given that $|1-a| + \sqrt{c-4} = 1$ and $b+d = 1$, find the value of xy. (Here $\lfloor r\rfloor$ represents the greatest integer not exceeding real number r, and $\{r\} = r - \lfloor r\rfloor$.)

Problem 1.26 Find the product of the positive roots of $\sqrt{2000x^{\log_{2000}x}} = x^2$

Problem 1.27 Solve the equation for x (lg means log base 10):

$$2^{\lg x} \cdot x^{\lg 2} - 3x^{\lg 2} - 2^{1+\lg x} + 4 = 0.$$

Problem 1.28 Let a be a real number. The quadratic equation

$$x^2 - 2x + \log_a(a^2 - a) = 0$$

has one positive root and one negative root. What is the range of values for the number a?

Problem 1.29 Find all the solutions to the system of equations

$$\begin{aligned}
\log_{10}(2000xy) - (\log_{10}x)(\log_{10}y) &= 4 \\
\log_{10}(2yz) - (\log_{10}y)(\log_{10}z) &= 1 \\
\log_{10}(zx) - (\log_{10}z)(\log_{10}x) &= 0
\end{aligned}$$

Problem 1.30 Determine the value of xy if $\log_8 x + \log_{16} y^2 = 2$ and $\log_8 y + \log_{16} x^2 = 3$.

Problem 1.31 Solve the inequality

$$\frac{1}{\log_2(x-1)} < \frac{1}{\log_2\sqrt{x+1}}$$

Problem 1.32 Let n be an even positive integer, solve the inequality for x:

$$\log_2 x - 4\log_{2^2} x + 12\log_{2^3} x + \cdots + n(-2)^{n-1}\log_{2^n} x > \frac{1-(-2)^n}{3}\log_2(x^2-2)$$

Problem 1.33 The function $y = \dfrac{kx+7}{kx^2+4kx+3} - \log_{1/3}\left(\dfrac{1}{4}x^2 - \sqrt{2}kx - 5k + 3\right)$ is defined on all $x \in \mathbb{R}$, find the possible range of k.

Problem 1.34 Solve for x: $|\lg x| > |\lg 4x| > |\lg 2x|$. ($\lg$ means log with base 10.)

Problem 1.35 Let A be an acute angle. Given that $\ln(1+\sin A) = a$ and $\ln \dfrac{1}{1-\sin A} = b$, compute $\ln \cos A$.

2. Fundamentals of Complex Numbers

Fundamentals of Complex Numbers

A *complex number* is a number consisting of a real number and an imaginary number. It can be written as $z = a + bi$, where a and b are real numbers, and i is the imaginary unit where $i^2 = -1$. The real numbers are contained in the complex numbers. The following theorem illustrates a very important difference between complex numbers and real numbers.

Theorem 2.1 Fundamental Theorem of Algebra

Every nonconstant polynomial with complex coefficients has at least one complex zero. Consequently, the number of zeros of a polynomial equals the degree, multiplicities counted.

This theorem is also known as the Gauss-d'Alembert Theorem. In the theorem, a "zero" means one value of the variable that makes the value of polynomial equal 0, in other words, a root of the polynomial equation. Polynomial equations with complex coefficients always have solutions in the complex numbers. In contrast, polynomial equations with real coefficients do not always have real number solutions.

Rectangular form, polar form, and exponential form

A complex number $z = a + bi$ can be represented by a vector from the Origin $(0,0)$ to the point (a,b). The expression $a + bi$ is the *rectangular* form. The real number a is the *real part* of z, often denoted $\Re(z) = a$, and the real number b is the *imaginary part* of z, often denoted $\Im(z)$. The length of the vector, which is the distance from the Origin to the point (a,b), is called the *modulus*, denoted $|z| = \sqrt{a^2 + b^2}$. The angle between the vector and the positive x-axis is called the *argument*, denoted $\arg(z)$.

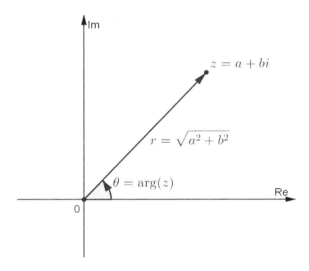

- A complex number can be represented using the modulus r and the argument θ, as follows. This is called the *polar form*:

$$z = r(\cos\theta + i\sin\theta)$$

where $a = r\cos(\theta)$ and $b = r\sin\theta$. Thus, $r = \sqrt{a^2 + b^2}$, and $\tan\theta = \dfrac{b}{a}$.
- Another form to represent a complex number is the *exponential form*:

$$z = re^{i\theta}.$$

Here e is Euler's number: $e = 2.718281828459045\ldots$.
- Sometimes we write the polar form in a shortened way:

$$r(\cos\theta + i\sin\theta) = r\operatorname{cis}\theta.$$

- Euler's identity:

$$e^{i\theta} = \cos\theta + i\sin\theta.$$

Rules of operations

The arithmetic operations on complex numbers are defined accordingly. Let a, b, c, d be real numbers, then

- Addition and subtraction:

$$(a + bi) \pm (c + di) = (a \pm c) + (b \pm d)i$$

- Multiplication:

$$(a + bi)(c + di) = (ac - bd) + (bc + ad)i$$

- Division:

$$\frac{a + bi}{c + di} = \frac{(a + bi)(c - di)}{(c + di)(c - di)} = \frac{ac - bd}{c^2 + d^2} + \frac{bc + ad}{c^2 + d^2}i$$

Complex conjugates

Let a and b be real numbers and $z = a + bi$, then the *complex conjugate* of z is $\bar{z} = a - bi$. The complex conjugates have the following properties:

- $\bar{\bar{z}} = z$
- The sum and product of z and $\bar{z}$ are real numbers:

$$z + \bar{z} = 2a, \quad z \cdot \bar{z} = a^2 + b^2 = |z|^2.$$

- A complex number z is a real number if and only if $z = \bar{z}$.
- Let z_1 and z_2 be complex numbers, then

$$
\begin{aligned}
\overline{z_1 + z_2} &= \overline{z_1} + \overline{z_2}, \\
\overline{z_1 - z_2} &= \overline{z_1} - \overline{z_2}, \\
\overline{z_1 \cdot z_2} &= \overline{z_1} \cdot \overline{z_2}, \\
\overline{z_1 / z_2} &= \overline{z_1} / \overline{z_2}.
\end{aligned}
$$

More properties

- Let $z_1 = r_1(\cos\theta_1 + i\sin\theta_1)$, $z_2 = r_2(\cos\theta_2 + i\sin\theta_2)$, then

$$z_1 \cdot z_2 = r_1 r_2(\cos(\theta_1 + \theta_2) + i\sin(\theta_1 + \theta_2)),$$

$$\frac{z_1}{z_2} = \frac{r_1}{r_2}(\cos(\theta_1 - \theta_2) + i\sin(\theta_1 - \theta_2))$$

In other words, when we multiply two complex numbers, we multiply the moduli and add the arguments (angles). When we divide two complex numbers, we calculate the ratio of the moduli and difference of the arguments.

- De Moivre's Theorem:

$$(\cos\theta + i\sin\theta)^n = \cos n\theta + i\sin n\theta;$$

written in the shortened form:

$$(\operatorname{cis}\theta)^n = \operatorname{cis} n\theta.$$

This theorem can be proved using mathematical induction.

Roots of Unity

Let n be a positive integer, we solve the following equation:

$$z^n = 1.$$

Let the root $z = r\operatorname{cis}\theta$, then according to De Moivre's Theorem,

$$1 = z^n = r^n(\operatorname{cis}\theta)^n = r^n\operatorname{cis} n\theta.$$

It is easy to see that $r = 1$. Since $1 = \cos 2k\pi + i\sin 2k\pi$ for any integer k,

$$\theta = \frac{2k\pi}{n}, \quad k \in \mathbb{Z}.$$

Therefore, there are n distinct roots for z:

$$z = \cos\frac{2k\pi}{n} + i\sin\frac{2k\pi}{n}, \quad k = 0, 1, \ldots, n-1.$$

The roots above are called the nth *roots of unity.*

Some special cases:

- When $n = 2$, the "square root of unity" is simply 1 and -1.
- When $n = 3$, the cube roots of unity are $1, \omega, \omega^2$, where

$$\omega = -\frac{1}{2} + \frac{\sqrt{3}}{2}i \quad \text{or} \quad \omega = -\frac{1}{2} - \frac{\sqrt{3}}{2}i.$$

- When $n = 4$, the 4th roots of unity are ± 1 and $\pm i$.

In general, the nth roots of unity are evenly distributed on the unit circle (the circle centered at the Origin with radius 1), including the number 1.

Let z be an nth root of unity, then the reciprocal of z is its complex conjugate, and is also an nth root of unity:

$$\frac{1}{z} = z^{-1} = z^{n-1} = \overline{z}.$$

Roots of any complex numbers

Let $a = r(\cos\theta + i\sin\theta)$ be a complex number. Then the equation

$$z^n = a$$

has the following roots:

$$z = \sqrt[n]{r}\left(\cos\frac{\theta + 2k\pi}{n} + i\sin\frac{\theta + 2k\pi}{n}\right), \quad k = 0, 1, 2, \ldots, n-1.$$

Let b represent one of the roots above, and define $\varepsilon = \cos\frac{2\pi}{n} + i\sin\frac{2\pi}{n}$, then all the roots above can be expressed as follows:

$$b, \ b\varepsilon, \ b\varepsilon^2, \ \ldots, \ b\varepsilon^{n-1}.$$

Complex numbers and geometry

There are many applications of complex numbers in solving geometry problems. Here are some of the basic properties:

- Let (a, b) and (c, d) be two points on the complex plane, and $z_1 = a + bi$ and $z_2 = c + di$ are the two corresponding complex numbers, then the midpoint of the segment connecting the two points is represented by $\frac{z_1 + z_2}{2}$.
- Given $\triangle ABC$ on the complex plane, let z_1, z_2, z_3 be the complex numbers representing the vertices A, B, C respectively. Then the centroid of $\triangle ABC$ is $\frac{z_1 + z_2 + z_3}{3}$.

2.1 Example Questions

Problem 2.1 Convert between the rectangular, polar, and exponential forms: write each of these numbers in the other forms.

(a) i

(b) $1 + i$

(c) $4\left(\cos\dfrac{\pi}{6} + i\sin\dfrac{\pi}{6}\right)$

Problem 2.2 Convert complex numbers among different forms:

(a) Let a and b be real numbers, convert $a + bi$ to polar form.

(b) Convert $re^{i\theta}$ to rectangular form.

Problem 2.3 What are the sets of points satisfying the following? Draw the diagrams.

(a) $|z| \le 2$

(b) $\Re z > \dfrac{1}{2}$ (The real part of z is greater than $\dfrac{1}{2}$)

(c) $\Re z = \Im z$

Problem 2.4 $|z_1 + z_2|^2 + |z_1 - z_2|^2 = 2\left(|z_1|^2 + |z_2|^2\right)$. What is the geometrical interpretation of this identity?

Problem 2.5 If $|a| = |b| = 1$ and $a + b + 1 = 0$, what are a and b?

Problem 2.6 Find the square roots of i.

Problem 2.7 If $f(z) = z^{3m+1} + z^{3m+2} + 1$. Show that $f(z)$ is divisible by $z^2 + z + 1$.

Problem 2.8 Let a and b be real numbers. Given that $2 + ai$ and $b + i$ are the two roots of the quadratic equation $x^2 + px + q = 0$ where p and q are real numbers. What are p and q?

Problem 2.9 Find the remainder when $x^{1001} - 1$ is divided by $x^4 + x^3 + 2x^2 + x + 1$.

Problem 2.10 Factor the polynomial: $x^8 + x^6 + x^4 + x^2 + 1$.

Problem 2.11 Use complex numbers to prove the following theorem:
In quadrilateral $ABCD$, let E, F, G, H be the midpoints of $\overline{AB}, \overline{BC}, \overline{CD}, \overline{DA}$ respectively, and let M and N be the midpoints of $\overline{AC}$ and $\overline{BD}$ respectively, then $\overline{EG}, \overline{FH}$ and $\overline{MN}$ are concurrent.

Problem 2.12 Let $\varepsilon = \cos\dfrac{2\pi}{n} + i\sin\dfrac{2\pi}{n}$, evaluate

$$(1 - \varepsilon)(1 - \varepsilon^2)\cdots(1 - \varepsilon^{n-1})$$

Problem 2.13 On the complex plane, given points $B(1), C(2+i)$, ABC is an equilateral triangle. Find the possible positions for A.

Problem 2.14 Evaluate: $1 + \dbinom{n}{3} + \dbinom{n}{6} + \cdots + \dbinom{n}{3m}$ where $3m$ is the maximum multiple of 3 not exceeding n.

Problem 2.15 Simplify the sums: $\displaystyle\sum_{k=0}^{n-1} \cos k\theta$ and $\displaystyle\sum_{k=0}^{n-1} \sin k\theta$

2.2 Practice Questions

Problem 2.16 Find the modulus of the complex number $z = 20 + 21i$.

Problem 2.17 Find the argument of the complex number $z = -2 - 2\sqrt{3}i$, in degrees between 0 and 360.

Problem 2.18 Assume $z^7 = 1$ and $z \neq 1$. Evaluate the following:

$$z^3 + \frac{1}{z^3} + z^6 + \frac{1}{z^6} + z^9 + \frac{1}{z^9}.$$

Problem 2.19 z is a complex number such that $z + \dfrac{1}{z} = 2\cos 3°$. Find the value of $z^{2000} + \dfrac{1}{z^{2000}} + 1$.

Problem 2.20 Let ω be the imaginary root of degree 3. Calculate

$$(1 - \omega)(1 - \omega^2)(1 - \omega^4)(1 - \omega^8).$$

Problem 2.21 What are the sets of points satisfying the following? Draw the diagrams.

(a) $2 < |z| \leq 4$

(b) $\Im z \le \dfrac{1}{2}$ (The imaginary part of z is less than or equal to $\dfrac{1}{2}$)

(c) $\left| \dfrac{z-1}{z+1} \right| < 1$

Problem 2.22 Let ω be an imaginary cube root of unity. Calculate

$$(1 - \omega + \omega^2)(1 - \omega^2 + \omega).$$

Problem 2.23 Given that $x^2 + x + 1 = 0$, calculate

$$\left(x + \frac{1}{x} \right)^2 + \left(x^2 + \frac{1}{x^2} \right)^2 + \cdots + \left(x^{27} + \frac{1}{x^{27}} \right)^2.$$

Problem 2.24 Let n be any positive integer, and $f(x) = x^{n+2} + (x+1)^{2n+1}$. Show that for any integer k, $f(k)$ is divisible by $k^2 + k + 1$.

Problem 2.25 Factor the polynomial: $x^{12} + x^9 + x^6 + x^3 + 1$.

Problem 2.26 Find the remainder when $x^{1234} + x^{2341} + x^{3412} + x^{4123}$ is divided by $x^4 + x^3 + x^2 + x + 1$.

Problem 2.27 Show that $x^{1979} + x^{1989} + x^{1999}$ is divisible by $x^7 + x^8 + x^9$.

Problem 2.28 Let $P(x)$, $Q(x)$. and $S(x)$ be polynomials satisfying

$$P(x^3) + xQ(x^3) = (x^2 + x + 1)S(x),$$

show that $x - 1$ is a factor of each of $P(x)$, $Q(x)$, and $S(x)$.

Problem 2.29 Solve for z:

$$\left(z + \frac{1}{z}\right)^2 + \left(z^2 + \frac{1}{z^2}\right)^2 - \left(z + \frac{1}{z}\right) \cdot \left(z^2 + \frac{1}{z^2}\right) \cdot \left(z^4 + \frac{1}{z^4}\right) = 3.$$

Problem 2.30 The solutions of the equation $z^4 + 4z^3 i - 6z^2 - 4zi - i = 0$ are the vertices of a convex polygon in the complex plane. What is the area of the polygon?

Problem 2.31 The region R in the complex plane contains all points z such that both $z/20$ and $20/\bar{z}$ have real and imaginary parts between 0 and 1, inclusive. Find the area of R.

Problem 2.32 Calculate the sums: $\displaystyle\sum_{k=1}^{n-1} k \cos \frac{2k\pi}{n}$ and $\displaystyle\sum_{k=1}^{n-1} k \sin \frac{2k\pi}{n}$

Problem 2.33 Find the sum: $\dbinom{n}{1} + \dbinom{n}{4} + \dbinom{n}{7} + \cdots$.

Problem 2.34 Find the sum: $\dbinom{n}{2} + \dbinom{n}{5} + \dbinom{n}{8} + \cdots$.

Problem 2.35 Find the following products:

(a) Find the product:

$$\prod_{k=1}^{n-1} \sin \frac{k\pi}{2n}.$$

(b) Find the product:

$$\prod_{k=1}^{n} \sin \frac{k\pi}{2n+1}.$$

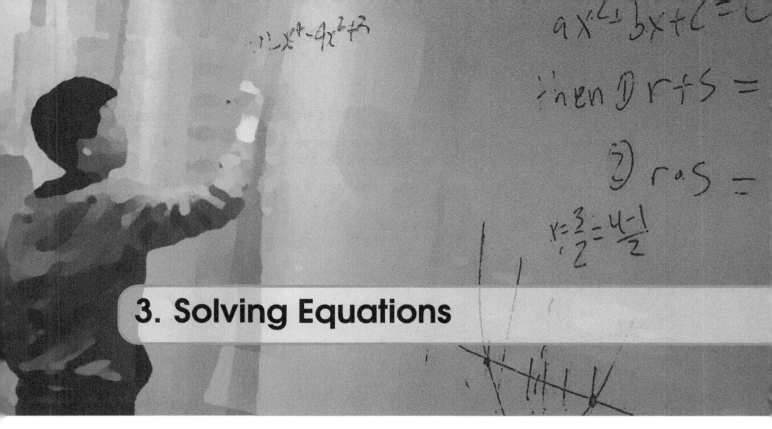

Useful Theorems

The following theorem is also known as the Gauss-d'Alembert Theorem. Its proof is beyond the scope of this class, so it is not included.

Theorem 3.1 Fundamental Theorem of Algebra

Every nonconstant polynomial with complex coefficients has at least one complex zero. Consequently, the number of zeros of a polynomial equals the degree, multiplicities counted.

Polynomial Division Algorithm

Let $P(x)$ be a polynomial, and $D(x)$ be a polynomial not identically 0, then there exist a unique pair of polynomials $Q(x)$ and $R(x)$ satisfying

$$P(x) = D(x)Q(x) + R(x),$$

where $\deg R < \deg D$.

Theorem 3.2 Polynomial Remainder Theorem

The remainder of a polynomial $P(x)$ divided by a linear divisor $x - \alpha$ is equal to $P(\alpha)$.

Proof. In the Division Algorithm, let $D(x) = x - \alpha$. Then

$$P(x) = (x - \alpha) \cdot Q(x) + r,$$

where r is a constant. Letting $x = \alpha$ on both sides, then we have

$$P(\alpha) = r,$$

which is to be proven.   ∎

Theorem 3.3 Factor Theorem

A polynomial $P(x)$ has a factor $x - \alpha$ if and only if $P(\alpha) = 0$.

Proof. This is a direct consequence of the Polynomial Remainder Theorem where $r = 0$.   ∎

Theorem 3.4 Vieta Theorem (quadratic version)

Let x_1, x_2 be the roots of a quadratic equation $ax^2 + bx + c = 0$, where $a \neq 0$, then

$$x_1 + x_2 = -\frac{b}{a}, \qquad \text{and} \qquad x_1 x_2 = \frac{c}{a}.$$

Proof. According to the Quadratic Formula,

$$x_{1,2} = \frac{-b \pm \sqrt{b^2 - 4ac}}{2a},$$

thus
$$x_1 + x_2 = \frac{-b + \sqrt{b^2 - 4ac}}{2a} + \frac{-b - \sqrt{b^2 - 4ac}}{2a} = \frac{-2b}{2a} = -\frac{b}{a},$$

and
$$x_1 \cdot x_2 = \frac{-b + \sqrt{b^2 - 4ac}}{2a} \cdot \frac{-b - \sqrt{b^2 - 4ac}}{2a} = \frac{b^2 - (b^2 - 4ac)}{4a^2} = \frac{4ac}{4a^2} = \frac{c}{a}.$$

∎

Theorem 3.5 Vieta Theorem (general version)

Let $P(x) = a_n x^n + a_{n-1} x^{n-1} + \cdots + a_0$ be a polynomial of degree n, and $x_1, x_2, \ldots, x_n$ be the zeros of $P(x)$. Then

$$\begin{cases} x_1 + x_2 + \cdots + x_n = -\dfrac{a_{n-1}}{a_n}, \\ x_1 x_2 + x_1 x_3 + \cdots + x_{n-1} x_n = \dfrac{a_{n-2}}{a_n}, \\ \cdots \\ x_1 x_2 \cdots x_n = (-1)^n \dfrac{a_0}{a_n}. \end{cases}$$

Proof. According to the Factor Theorem, $x - x_1, x - x_2, \ldots, x - x_n$ are all factors of $P(x)$. Therefore

$$a_n x^n + a_{n-1} x^{n-1} + \cdots + a_0 = a_n(x - x_1)(x - x_2) \cdots (x - x_n).$$

The desired relations are obtained by expanding the right hand side and comparing the coefficients. ∎

The following theorem is often useful when the coefficients of a polynomial are all integers. The proof of this theorem is straight forward.

> **Theorem 3.6 Rational Root Theorem**
>
> Let $P(x) = a_n x^n + a_{n-1} x^{n-1} + \cdots + a_0$ be a polynomial of degree n where all its coefficients $a_0, a_1, \ldots, a_n$ are integers, and assume a rational number $\dfrac{p}{q}$ is a root of the equation $P(x) = 0$, where $\gcd(p, q) = 1$, then $p \mid a_0$ and $q \mid a_n$.

3.1 Example Questions

Problem 3.1 Let $x = 100$ be a root of quadratic equation $ax^2 + bx + a = 0$. What is the other root?

Problem 3.2 The equation in x: $x^2 + px + q = 0$ has two nonzero integer roots, and $p + q = 198$. What is p?

Problem 3.3 Find real solutions for

$$\begin{cases} x + y & = 6 \\ xy - z^2 & = 9 \end{cases}$$

Problem 3.4 $(x^2 + x + 1)(x^2 + x + 2) = 12$

Problem 3.5 $(x^2 + x - 2)^2 + (2x^2 - 5x + 3)^2 = (3x^2 - 4x + 1)^2$

Problem 3.6 Find the real solutions: $x^2 + x + 1 = \dfrac{2}{x^2 + x}$

Problem 3.7 Find the real roots: $x^4 + (x - 2)^4 = 706$.

Problem 3.8 Find the real roots: $2x^4 - 9x^3 + 14x^2 - 9x + 2 = 0$.

Problem 3.9 Let x, y be **positive integers**, and satisfy

$$\begin{cases} xy + x + y &= 71 \\ x^2 y + xy^2 &= 880 \end{cases}$$

Find the value of $x^2 + y^2$.

Problem 3.10 Find the real solutions:

$$\begin{cases} x^3 + y^3 + x^3 y^3 &= 17 \\ x + y + xy &= 5 \end{cases}$$

Problem 3.11 Find the real solutions: $5x^2 + x - x\sqrt{5x^2 - 1} - 2 = 0$

Problem 3.12 Find the real solutions: $\dfrac{x-1}{x+1} + \dfrac{x-4}{x+4} = \dfrac{x-2}{x+2} + \dfrac{x-3}{x+3}$

Problem 3.13 Solve for x: $\left(\sqrt[5]{3}\right)^x + \left(\sqrt[10]{3}\right)^{x-10} = 84$

Problem 3.14 Find the real roots: $\sqrt{5 - \sqrt{5 - x}} = x$.

Problem 3.15 Solve for x: $(x - \sqrt{3})x(x+1) + 3 - \sqrt{3} = 0$.

3.2 Practice Questions

Problem 3.16 Let $ax^2 + bx + c = 0$ be a quadratic equation in x. Assume $b = a + c$, what number is guaranteed to be a root of this equation?

Problem 3.17 Assume $a^2 + 3a + 1 = 0$ and $b^2 + 3b + 1 = 0$, and $a \neq b$. Find the value of $\dfrac{a}{b} + \dfrac{b}{a}$.

Problem 3.18 The equation $(x - a)(x - 8) - 1 = 0$ has two integer roots. What is a?

Problem 3.19 If p, q are positive integers, and the equation (in x) $\dfrac{1}{2}px^2 - \dfrac{1}{2}qx + 997 = 0$ has two prime roots. Find the value of $2p + q$.

Problem 3.20 Solve for x: $4^x - 10 \cdot 2^{x-1} - 24 = 0$

Problem 3.21 Assume that m and n are real numbers and the quadratic equation (in x) $x^2 + 2(m + 1)x + (3m^2 + 4mn + 4n^2 + 2) = 0$ has real roots. Evaluate $3m^2 + 2n^2$.

Problem 3.22 Find the real solutions: $(2x^2 - 3x + 1)^2 = 22x^2 - 33x + 1$

Problem 3.23 $2x^4 + 3x^3 - 16x^2 + 3x + 2 = 0$.

Problem 3.24 Find the real roots of $\sqrt[3]{5-x}+\sqrt{x-4}=1$.

Problem 3.25 Find the real solutions:

$$(x^2+x+1)+(x^2+2x+3)+(x^2+3x+5)+\cdots+(x^2+20x+39)=4500$$

Problem 3.26 Find the real roots of

$$x+\sqrt{x+\frac{1}{2}+\sqrt{x+\frac{1}{4}}}=4.$$

Problem 3.27 Solve for x: $2^{x^2-1}-3^{x^2}=3^{x^2-1}-2^{x^2+2}$

Problem 3.28 The equation $|x^2-5x|=a$ has exactly two distinct real roots. What is the possible range of values for a?

Problem 3.29 Given an equation in x: $x^2+(m-2)x+\frac{1}{2}m-3=0$.

(a) Show that no matter what real value m is, this equation always has two distinct real roots.

(b) Let x_1, x_2 be the two real roots of the equation, satisfying $2x_1 + x_2 = m + 1$, find the possible values of m.

Problem 3.30 Solve the equation for x:

$$\sqrt{4 + \sqrt{4 - x}} = x.$$

Problem 3.31 Let a be a real number, and the equation $x^2 + a^2x + a = 0$ has real roots for x. Find the maximum possible root x.

Problem 3.32 Solve for x: $\sqrt[3]{(x+1)^2} + 4\sqrt[3]{(x-1)^2} = 5\sqrt[3]{x^2 - 1}$

Problem 3.33 $\sqrt[3]{x} + \sqrt[3]{2 - x} = 2$

Problem 3.34 Find the real solutions:

$$\begin{cases} x^3 + y^3 - 18xy = 0 \\ x^2 + y^2 - 20x = 0 \end{cases}$$

Problem 3.35 Find the real solutions for

$$\frac{(39-x)\sqrt[5]{x-6} - (x-6)\sqrt[5]{39-x}}{\sqrt[5]{39-x} - \sqrt[5]{x-6}} = 30.$$

4. Solving Inequalities

Important Inequalities

Trivial Inequalities

The following are trivial inequalities.

- $|x| \geq 0$ for all real numbers x. (This is true for complex numbers as well.)
- $x^2 \geq 0$ for all real numbers x.
- $\sqrt{x} \geq 0$ for all $x \geq 0$.

Triangle Inequality

- Let a and b be real numbers, then

$$|a+b| \leq |a| + |b|.$$

In fact the triangle inequality is true for all complex numbers.
- Let a and b be real numbers, then

$$||a| - |b|| \leq |a+b|.$$

This is a direct consequence of the Triangle Inequality above.
- Let $a_1, a_2, \ldots, a_n$ be real numbers, then

$$|a_1 + a_2 + \cdots + a_n| \leq |a_1| + |a_2| + \cdots + |a_n|.$$

This is another direct consequence of the Triangle Inequality.

Some Common Inequalities

The following are some common inequalities. To solve problem related to inequalities, it is important to know AM-GM-HM, Cauchy-Schwarz, and Rearrangement Inequalities. There are multiple ways to prove each of these inequalities. One proof is provided for each of them in this chapter, and it is strongly encouraged that you find other proofs and compare the different proofs.

Theorem 4.1 AM-GM-HM

Let $a_1, a_2, \ldots, a_n$ be positive real numbers. Then

$$\frac{a_1 + a_2 + \cdots + a_n}{n} \geq \sqrt[n]{a_1 a_2 \cdots a_n} \geq \frac{n}{\dfrac{1}{a_1} + \dfrac{1}{a_2} + \cdots + \dfrac{1}{a_n}},$$

where equality holds if and only if $a_1 = a_2 = \cdots = a_n$.

Proof. We only need to prove the AM-GM part. The GM-HM part is a direct consequence of the AM-GM part by applying AM-GM on $\dfrac{1}{a_1}, \dfrac{1}{a_2}, \ldots, \dfrac{1}{a_n}$.

To show the AM-GM part, let

$$A = \frac{a_1 + a_2 + \cdots + a_n}{n}, \quad G = \sqrt[n]{a_1 a_2 \cdots a_n}.$$

We apply a "smoothing" technique. If $a_1 = a_2 = \cdots = a_n$, then $A = G = a_1$ and we are done.

If the a_i's are not all the same, then one of the a_i's, say a_k ($1 \leq k \leq n$), is less than A. Otherwise, $a_i \geq A$ for all $1 \leq i \leq n$, and at least one of the inequalities is strict, then $\dfrac{a_1 + a_2 + \cdots + a_n}{n} > A$, contradicts the definition of A.

Similarly, there is some a_j ($1 \leq j \leq n$) such that $a_j > A$. Thus $a_k < A < a_j$.

Replace a_k and a_j with a_k' and a_j', where $a_k' = A$, and $a_j' = a_k + a_j - A$. Then $a_k' + a_j' = a_k + a_j$, so the Arithmetic Mean A is unchanged.

On the other hand,

$$
\begin{aligned}
a'_k a'_j - a_k a_j &= A(a_k + a_j - A) - a_k a_j \\
&= A \cdot a_k + A \cdot a_j - A^2 - a_k a_j \\
&= (a_j - A)(A - a_k) \\
&> 0,
\end{aligned}
$$

therefore the Geometric Mean G is increased after the replacement.

After the replacement, if the numbers a_i are still not all the same, we can repeat the process above. At each step, the Arithmetic Mean (AM) is never changed, and one of the values is changed to equal the AM; the Geometric Mean strictly increases after each step. After at most n steps, all the values will be changed to equal A, and the GM has increased to equal A.

Therefore,

$$
\frac{a_1 + a_2 + \cdots + a_n}{n} \geq \sqrt[n]{a_1 a_2 \cdots a_n},
$$

and equality occurs if and only if $a_1 = a_2 = \cdots = a_n$. ∎

Theorem 4.2 Cauchy-Schwarz

For any real numbers $a_1, a_2, \ldots, a_n, b_1, \ldots, b_n$,

$$
(a_1^2 + a_2^2 + \cdots + a_n^2)(b_1^2 + b_2^2 + \cdots + b_n^2) \geq (a_1 b_1 + a_2 b_2 + \cdots + a_n b_n)^2,
$$

where equality holds if the two sequences are proportional.

Proof. Let $A = \displaystyle\sum_{i=1}^{n} a_i^2$, $B = 2\displaystyle\sum_{i=1}^{n} a_i b_i$, $C = \displaystyle\sum_{i=1}^{n} b_i^2$. Consider the quadratic polynomial

$$
Ax^2 + Bx + C = \sum_{i=1}^{n}(a_i^2 x^2 + 2a_i b_i x + b_i^2) = \sum_{i=1}^{n}(a_i x + b_i)^2 \geq 0.
$$

The polynomial is nonnegative for all $x \in \mathbb{R}$, thus it has at most one real root, which means the discriminant $B^2 - 4AC \leq 0$:

$$
B^2 - 4AC = 4\left(\sum_{i=1}^{n} a_i b_i\right)^2 - 4\left(\sum_{i=1}^{n} a_i^2\right)\left(\sum_{i=1}^{n} b_i^2\right) \leq 0,
$$

Canceling the number 4, this is exactly the inequality to be proven. ∎

Theorem 4.3　Rearrangement Inequality

Let $a_1, a_2, \ldots, a_n$ and $b_1, b_2, \ldots, b_n$ be real numbers. Then the sum

$$a_1 b_1 + a_2 b_2 + \cdots + a_n b_n$$

is maximized when the a_i's and b_i's are ordered in the same way (i.e. $a_1 \geq a_2 \geq \cdots \geq a_n$ and $b_1 \geq b_2 \geq \cdots \geq b_n$) and is minimized when they are ordered in the opposite way (i.e. $a_1 \geq a_2 \geq \cdots \geq a_n$ and $b_1 \leq b_2 \leq \cdots \leq b_n$).

Proof. Without loss of generality, assume $a_1 \geq a_2 \geq \cdots \geq a_n$.

We prove that the given sum is maximized when $b_1 \geq b_2 \geq \cdots \geq b_n$. Assume some pair of the b_i's, say b_k and b_l where $1 \leq k < l \leq n$, is in the opposite order, i.e. $b_k \leq b_l$. Since $a_k \geq a_l$, we have

$$(a_k b_l + a_l b_k) - (a_k b_k + a_l b_l) = (a_k - a_l)(b_l - b_k) \geq 0,$$

that is, "swapping" b_k and b_l increases the sum. There are only finite number of pairs of b_i's that are in the opposite order, and each "swapping" reduces the number by at least 1. Therefore, after finite steps of the above "swapping", the a_i's and b_i's are ordered in the same way, and the sum is maximized.

On the other hand, suppose the a_i's and b_i's are ordered in the opposite way, define $c_i = -b_i$ for $i = 1, 2, \ldots, n$, then the a_i's and the c_i's are ordered in the same way. Therefore the sum

$$a_1 c_1 + a_2 c_2 + \cdots + a_n c_n = -(a_1 b_1 + a_2 b_2 + \cdots + a_n b_n)$$

is maximized, and hence the sum $a_1 b_1 + a_2 b_2 + \cdots + a_n b_n$ is minimized.

■

Strategies for Solving Inequalities

The following is an incomplete list of common strategies when solving inequalities.

- Pay attention to domain of definitions. This includes (but not limited to):
  - Denominators cannot be 0.

- Roots of even index, i.e. $\sqrt[n]{A}$ where n is even and A is an algebraic expression: $A \geq 0$.
- Logarithms: In $\log_A B$ where A and B are algebraic expressions, it is required that $A > 0$, $A \neq 1$, and $B > 0$.
- For problems related to absolute value $|A|$ where A is an algebraic expression, analyze the cases where $A \geq 0$ and $A < 0$.
- Apply the common inequalities.
  - If the sum of several positive numbers is a constant value S, then according to the AM-GM, their product is maximized when all the numbers are equal.
  - If the product of several positive numbers is a constant value P, then their sum is minimized when all the numbers are equal.
- Change of Variables: a change of variables is a basic technique where quantities in a mathematical expression are replaced with new variables or functions of new variables. Usually, after the change of variables, the problem becomes simpler or equivalent to a better understood problem.
  - Repeated expressions: if certain expression appears multiple times in a problem, the problem is usually simplified if this expression is replaced with a new variable.
  - Radicals: using a new variable to represent a complicated radical expression is often a good idea.
  - Trigonometric substitution: for expressions like $\sqrt{1-x^2}$, one of the best ways is $x = \sin\theta$, then $\sqrt{1-x^2} = \cos\theta$ for properly chosen range for θ.
  - Another trigonometric substitution: for expressions containing $1+x^2$: let $x = \tan\theta$, then $1+x^2 = \sec^2\theta$.

4.1 Example Questions

Problem 4.1 Compare: find out which is bigger in each pair, without using a calculator.
(a) $\sqrt{3}$ and $2^{0.8}$

(b) 16^{18} and 18^{16}

(c) 2 and $\sqrt{2 + \sqrt{2 + \sqrt{2 + \sqrt{3}}}}$

Problem 4.2 Solve the inequality: $|-x^2 + 2x - 3| < |3x - 1|$

Problem 4.3 Solve the inequality: $|\log_{\sqrt{2}} x - 2| - |\log_2 x - 2| < 2$

Problem 4.4 Solve the inequality: $\dfrac{4x^2}{(1 - \sqrt{1 + 2x})^2} < 2x + 9$

Problem 4.5 Solve for x: $\dfrac{2}{3}(27^x - 12^x) > 18^x - 8^x$.

Problem 4.6 $\sqrt{x^2 + 4x} \leq 4 - \sqrt{16 - x^2}$

Problem 4.7 If x, y, z are positive and $x + y + z = 1$, find the minimum value of

$$\frac{1}{x} + \frac{4}{y} + \frac{9}{z}.$$

Problem 4.8 Assume $a, b, c > 0$, find the minimum values of the following:

$$\frac{a}{b+c} + \frac{b}{c+a} + \frac{c}{a+b}.$$

(**Hint:** Add 1 to each fraction, and apply Cauchy-Schwarz Inequality.)

Problem 4.9 Assume $0 < x < 1$, find the maximum value of $x(1 - x^4)$. (Reminder: calculus is not allowed.)

Problem 4.10 Let A, B, C be the 3 angles of $\triangle ABC$. What is the minimum value of $\frac{1}{A^2} + \frac{1}{B^2} + \frac{1}{C^2}$? (Angles are in radians)

Problem 4.11 Find the smallest positive value k such that

$$\log_{10}(xy) \leq \log_{10} k \cdot \sqrt{\log_{10}^2 x + \log_{10}^2 y}$$

is true for all $x > 1$ and $y > 1$.

Problem 4.12 Find the maximum and minimum of $\sin x + \cos x$.

Problem 4.13 Find the range of the function $y = \sqrt{4x - 1} + \sqrt{2 - x}$.

Problem 4.14 Solve for x: $\dfrac{x}{\sqrt{1+x^2}} + \dfrac{1-x^2}{1+x^2} > 0$.

Problem 4.15 If $2x + y \geq 1$, find the minimum value for $u = y^2 - 2y + x^2 + 4x$.

4.2 Practice Questions

Problem 4.16 Let x, y, z be positive real numbers, and $x + 3y + 4z = 6$, find the maximum value for $x^2 y^3 z$.

Problem 4.17 Assume $a, b, c > 0$, find the minimum values of the following:

$$\frac{a}{b+c} + \frac{4b}{c+a} + \frac{5c}{a+b}.$$

Problem 4.18 The inequality $-9 < \dfrac{3x^2 + px + 6}{x^2 - x + 1} \leq 6$ holds for all $x \in \mathbb{R}$. What is p?

Problem 4.19 Let n be positive integer. If the inequality

$$(x^2 + y^2 + z^2)^2 \leq n(x^4 + y^4 + z^4)$$

holds for all x, y, and z. What is the minimum value of n?

Problem 4.20 If $x^2 + y^2 \leq 2\sqrt{2}$, find the maximum value of $|x^2 - 2xy - y^2|$.

Problem 4.21 Compare: find out which is bigger in each pair, without using a calculator.
(a) $\sqrt[8]{8!}$ and $\sqrt[9]{9!}$

(b) 37^{73} and $73!$

Problem 4.22 Solve the inequality: $|x+7| - |3x-4| + \sqrt{3-2\sqrt{2}} > 0$

Problem 4.23 Solve the inequality: $x^{\log_4 x} > \dfrac{x^4 \sqrt{x}}{16}$.

Problem 4.24 The solution set for $\log_x(5x^2 - 8x + 3) > 2$ is a subset of the solution set for $x^2 - 2x - a^4 + 1 \geq 0$. Find the set of possible values for a.

Problem 4.25 Assume $a, b, c > 0$, find the minimum values of the following:

$$\frac{a}{b+3c} + \frac{b}{8c+4a} + \frac{9c}{3a+2b}.$$

Problem 4.26 In tetrahedron $P - ABC$, $\angle APB = \angle BPC = \angle CPA = 90°$, and the sum of the lengths of the six edges is S. Find the maximum possible value for its volume.

Problem 4.27 Find the maximum and minimum of $|\sin x| + |\cos x|$. (Hint: Square the expression.)

Problem 4.28 Find the maximum and minimum values of: $\sqrt{5-x^2} + \sqrt{3}x$

Problem 4.29 If the equation $2a \cdot 9^{\sin x} + 4a \cdot 3^{\sin x} + a - 8 = 0$ has real solution for x, find the set of possible values for a.

Problem 4.30 Find the minimum value of the function

$$g(x) = \sqrt{\sin^2 x - 2\sin x + \frac{17}{16}} + \sqrt{\sin^2 x - \frac{2}{3}\sin x + \frac{17}{144}}.$$

Problem 4.31 Let θ be an acute angle that satisfies

$$\frac{1}{\sin^6 \theta} + \frac{81}{\cos^6 \theta} = 256.$$

Find the range of possible values of θ in degrees.
(**Hint**: Use the fact $\sin^2 \theta + \cos^2 \theta = 1$, and apply the Cauchy-Schwarz Inequality.)

Problem 4.32 Equation $x^2 - mx + 4 = 0$ has at least one root for x in $(-1, 1)$. Find the set of possible values for parameter m.

Problem 4.33 Let n be a positive integer. Given $a_i > 0 (i = 1, 2, \ldots, n)$ satisfying $\sum_{i=1}^{n} a_i = 1$, find the minimum value of

$$S = \frac{a_1}{1 + a_2 + a_3 + \cdots + a_n} + \frac{a_2}{1 + a_1 + a_3 + \cdots + a_n} + \cdots + \frac{a_n}{1 + a_1 + a_2 + \cdots + a_{n-1}}.$$

Problem 4.34 Assume the sum of four numbers is 4, and the sum of their squares is 8. Find the maximum possible value for the largest of the four numbers.

Problem 4.35 Real numbers a, b, c satisfy $a + b = 2c - 3$, $a^2 + b^2 - 2c^2 + 8c = 5$. Find the minimum value for ab.

Problem 4.36 Let $f(x) = x^2 - x + k$, and $\log_2 f(a) = 2$, $f(\log_2 a) = k$, $a > 0, a \neq 1$, find the minimum value for $f(\log_2 x)$.

Problem 4.37 If $x, y, z \in \mathbb{R}^+$ and $x^4 + y^4 + z^4 = 1$, find the minimum value of

$$\frac{x^3}{1 - x^8} + \frac{y^3}{1 - y^8} + \frac{z^3}{1 - z^8}.$$

Problem 4.38 Let x, y, z, w be real numbers, not all 0, satisfying

$$xy + 2yz + zw \leq A(x^2 + y^2 + z^2 + w^2),$$

find the minimum value of A.

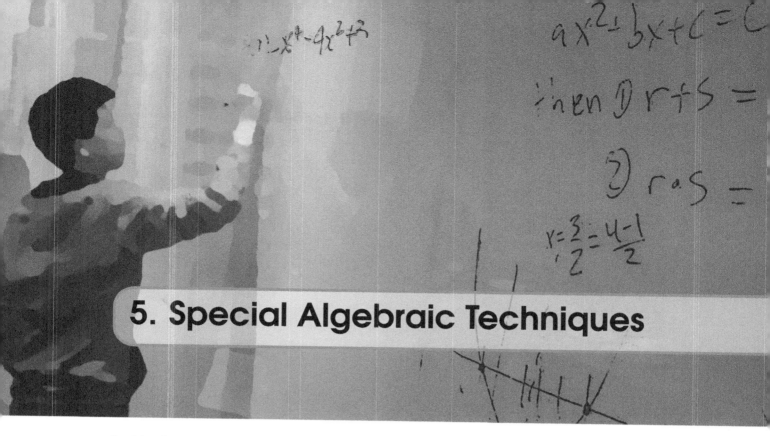

5. Special Algebraic Techniques

In this chapter, we review some special algebraic techniques and use them to solve problems in algebra, including solving equations or inequalities, and performing algebraic transformations.

Solving Equations by Observation

It is easy to observe and identify solutions. The drawback of this method is that you don't know whether you have all the solutions. Methods exist to resolve those issues.

Example 5.1

Find the real root: $x + \sqrt{x - 2} = 4 + \sqrt{2}$.

Solution

It is easy to see $x = 4$ is a solution by observation. To see this is the only root, note that the function $f(x) = x + \sqrt{x - 2}$ is strictly increasing for $x \geq 2$, hence there is a unique value of x for which $f(x)$ equals the right hand side.

Completing the Square

Completing the square is a very basic skill. It is widely useful in factoring, quadratic equations, quadratic functions and inequalities.

When solving equations, completing the square can help reducing the degree (taking square roots), and also finding situations where nonnegative values are forced to be 0 (using the fact that no squares are negative).

Example 5.2

Find real roots for

$$\sqrt{x - \frac{1}{x}} + \sqrt{1 - \frac{1}{x}} = x.$$

Solution

Clearly $x \geq 1$. Multiply both sides by 2 and complete the square,

$$0 = 2x - 2\sqrt{x - \frac{1}{x}} - 2\sqrt{1 - \frac{1}{x}}$$

$$= \left[\left(x - \frac{1}{x} \right) - 2\sqrt{x - \frac{1}{x}} + 1 \right] + \left[(x - 1) - 2\frac{\sqrt{x - 1}}{\sqrt{x}} + \frac{1}{x} \right]$$

$$= \left(\sqrt{x - \frac{1}{x}} - 1 \right)^2 + \left(\sqrt{x - 1} - \frac{1}{\sqrt{x}} \right)^2.$$

Now we have $\sqrt{x - \frac{1}{x}} = 1$ and $\sqrt{x - 1} = \frac{1}{\sqrt{x}}$. Both of them give the following equation:

$$x^2 - x - 1 = 0.$$

Solve it and take the root greater than or equal to 1, $x = \dfrac{1 + \sqrt{5}}{2}$. Checking with the equation, this is the solution.

Note: You can also square both sides (after rearranging the terms) before completing the square.

Change of Variables

The benefits of change of variables include eliminating variables, reducing powers, rationalizing, removing fractions, etc.

Example 5.3

Solve for x:

$$\frac{\sqrt{16+\sqrt{x}}}{16} + \frac{\sqrt{16+\sqrt{x}}}{\sqrt{x}} = \sqrt[4]{\frac{x}{16}}.$$

Solution

It is not easy to handle 4th roots or nested radicals, so we make the change of variables: $y = \sqrt{x}$. Then

$$\frac{\sqrt{16+y}}{16} + \frac{\sqrt{16+y}}{y} = \frac{\sqrt{y}}{2},$$

that is,

$$\frac{\sqrt{16+y}(16+y)}{16y} = \frac{\sqrt{y}}{2},$$

so

$$\sqrt{\left(\frac{16+y}{y}\right)^3} = 8,$$

therefore

$$\frac{16+y}{y} = 4,$$

and we get $y = \dfrac{16}{3}$, and $x = y^2 = \dfrac{256}{9}$. Verifying, it is a root.

Example 5.4

Solve for x:

$$x^3 + 2\sqrt{11}x^2 + 11x + \sqrt{11} + 1 = 0.$$

Solution

If you are not familiar with Cardano-Tartaglia formula for cubic equations, it would be very hard to solve for x directly. However, if we treat $\sqrt{11}$ as the "variable" and x as a parameter, we can solve this equation as a quadratic.

Let $y = \sqrt{11}$, then the equation becomes:

$$x^3 - 2yx^2 + y^2x + y + 1 = 0.$$

Rewrite the equation in terms of y, then the equation has $y = \sqrt{11}$ as a solution:

$$xy^2 + (2x^2 + 1)y + (x^3 + 1) = 0.$$

Since $x \neq 0$, apply the quadratic formula,

$$y = \frac{-(2x^2 + 1) \pm \sqrt{(2x^2 + 1)^2 - 4x(x^3 + 1)}}{2x}.$$

We first simplify the discriminant:

$$
\begin{aligned}
(2x^2 + 1)^2 - 4x(x^3 + 1) &= 4x^4 + 4x^2 + 1 - 4x^4 - 4x \\
&= 4x^2 - 4x + 1 \\
&= (2x + 1)^2,
\end{aligned}
$$

thus

$$y = \frac{-(2x^2 + 1) \pm (2x + 1)}{2x},$$

so simplify each case,

$$y = -x - 1, \quad y = \frac{-x^2 + x - 1}{x}.$$

Now solve for x, using the fact that $y = \sqrt{11}$,

$$x_1 = -1 - y = -1 - \sqrt{11},$$

and

$$x_{2,3} = \frac{1 - y \pm \sqrt{y^2 - 2y - 3}}{2} = \frac{1 - \sqrt{11} \pm \sqrt{8 - 2\sqrt{11}}}{2}.$$

Other Special Techniques

> **Example 5.5**
>
> Solve for x:
> $$\sqrt{\frac{3-x}{1+x}} = \frac{3-x^2}{x^2+1}.$$

Solution

To solve this equation, we use a new variable y, and convert the equation to a system of equations.

Let
$$y = \frac{3-x^2}{x^2+1} = \sqrt{\frac{3-x}{1+x}} \geq 0,$$

then
$$\begin{cases} yx^2 + x^2 + y - 3 = 0 \\ xy^2 + y^2 + x - 3 = 0. \end{cases}$$

Subtract, $(x-y)(xy+x+y-1) = 0$, so $x = y$ or $xy+x+y-1 = 0$.

Case 1: $x = y$, eliminating y from the first equation above, $(x-1)(x^2+2x+3) = 0$, so $x = 1$.

Case 2: $xy+x+y-1 = 0$, so $y = \dfrac{1-x}{1+x}$. From $y \geq 0$ we get $-1 < x \leq 1$. Substituting y into the first equation and simplify, $x^2 - 2x - 1 = 0$, therefore $x = 1 \pm \sqrt{2}$, but $1 + \sqrt{2}$ is greater than 1, so only $1 - \sqrt{2}$ is a root.

Checking, both 1 and $1 - \sqrt{2}$ are roots.

Example 5.6

Solve the system of equations:

$$\begin{cases} \left(\dfrac{1+\sqrt{3}}{2}\right)^2 x + \left(\dfrac{1+\sqrt{3}}{2}\right) y + 1 = 0, \\[4mm] \left(\dfrac{1-\sqrt{3}}{2}\right)^2 x + \left(\dfrac{1-\sqrt{3}}{2}\right) y + 1 = 0. \end{cases}$$

Solution

In this problem, the system of two equations can be seen as follows. The two values $\dfrac{1+\sqrt{3}}{2}$ and $\dfrac{1-\sqrt{3}}{2}$ are the roots of a quadratic equation in t:

$$xt^2 + yt + 1 = 0.$$

From Vieta's Theorem,

$$-\frac{y}{x} = \frac{1+\sqrt{3}}{2} + \frac{1-\sqrt{3}}{2} = 1,$$

and

$$\frac{1}{x} = \frac{1+\sqrt{3}}{2} \cdot \frac{1-\sqrt{3}}{2} = -\frac{1}{2}.$$

Therefore $x = -2, y = 2$.

5.1 Example Questions

Problem 5.1 Find one solution for $x^{x^2} = 2$ by observation.

Problem 5.2 Find the sum of the roots of $x^2 - 2000|x| = 2000$.

Problem 5.3 Solve for x, using the method of Completing the Square.

$$x + \frac{x}{\sqrt{x^2 - 1}} = \frac{35}{12}.$$

Hint: first square both sides.

Problem 5.4 Solve for x: $\sqrt{5 - x} + \sqrt{2 + x} = \sqrt{5} + \sqrt{2}$.

Problem 5.5 Solve for x:

$$(6x + 7)^2 (3x + 4)(x + 1) = 6.$$

Hint: Convert $3x + 4$ to $6x + 8$ and $x + 1$ to $6x + 6$.

Problem 5.6 Solve for x:

$$(3x^2 - 2x + 1)(3x^2 - 2x - 7) + 12 = 0.$$

Problem 5.7 Find real solutions:

$$\sqrt{x - 1} + 2\sqrt{y - 4} + 3\sqrt{z - 9} + 4\sqrt{w - 16} = \frac{1}{2}(x + y + z + w).$$

Problem 5.8 Find all real roots for $x^2 - 2x \sin \frac{\pi x}{2} + 1 = 0$.

Problem 5.9 Given $a > b$, solve for x:

$$\sqrt{a-x}+\sqrt{x-b}=\sqrt{a-b}.$$

Problem 5.10 Let real numbers $x_1, x_2, \ldots, x_n$ satisfy

$$\frac{x_1}{x_1^2+1}=\cdots=\frac{x_n}{x_n^2+1},$$
$$x_1+\cdots+x_n+\frac{1}{x_1}+\cdots+\frac{1}{x_n}=\frac{10}{3}.$$

Find the value of x_n.

Problem 5.11 Solve for x, y:

$$\begin{cases} \sqrt{3-y}=\sqrt{x}+\sqrt{x-y}, \\ \sqrt{1-y}=\sqrt{x}-\sqrt{x-y}. \end{cases}$$

Problem 5.12 (IMO 1959) Find all possible values of x satisfying the equation

$$\sqrt{x+\sqrt{2x-1}}+\sqrt{x-\sqrt{2x-1}}=\sqrt{2}.$$

Problem 5.13 Find real solutions to the system

$$\begin{cases} y = \sqrt{x - \dfrac{1}{x}} + \sqrt{1 - \dfrac{1}{x}}, \\ x = \sqrt{y - \dfrac{1}{y}} + \sqrt{1 - \dfrac{1}{y}}. \end{cases}$$

Hint: From the first equation, calculate $2x - 2y$, and complete the squares. Then obtain a similar result from the second equation.

Problem 5.14 Find **positive** solutions:

$$\begin{cases} x^2 + y^2 + xy = 1, \\ y^2 + z^2 + yz = 3, \\ z^2 + x^2 + zx = 4. \end{cases}$$

Hint: Is there any geometric interpretation?

Problem 5.15 Find all **integer roots** of $\dfrac{x+y}{x^2 - xy + y^2} = \dfrac{3}{7}$.

5.2 Practice Questions

Problem 5.16 Find real solutions:

$$x = \sqrt{2 + \sqrt{2 + \sqrt{2 + \sqrt{2 + x}}}}$$

Problem 5.17 Solve the equation:

$$\frac{x-7}{\sqrt{x-3}+2} + \frac{x-5}{\sqrt{x-4}+1} = \sqrt{10}.$$

Problem 5.18 Solve for x: $\sqrt{x+3-4\sqrt{x-1}} + \sqrt{x+8-6\sqrt{x-1}} = 1$.

Problem 5.19 $\sqrt{x+8} + \sqrt{x+3} = \sqrt{3x+6} + \sqrt{3x+1}$.

Problem 5.20 Solve for x:

$$\frac{1}{x^2+1} + \frac{x^2+1}{x^2} = \frac{10}{3x}.$$

Problem 5.21 Solve the system:

$$\begin{cases} x+y & = -20, \\ \sqrt[3]{x-1}+\sqrt[3]{y+2} & = -1. \end{cases}$$

Problem 5.22 Solve the equation:

$$\frac{x^3+2}{x^2-x+1}+\frac{x^3-2}{x^2+x+1}=2x.$$

Problem 5.23 Solve the equation:

$$\frac{2x}{3}=\frac{x^2}{12}+\frac{3}{x^2}+\frac{4}{x}.$$

Problem 5.24 Assume $a < 30$ is an integer, and the equation (in x)

$$\sqrt{2x-4}-\sqrt{x+a}=1$$

has exactly one integer root. Find all possible values for a.

Problem 5.25 (ZIML Varsity Feb 2018) Let x, y be real numbers satisfying

$$(x + \sqrt{x^2 - 2018})(y + \sqrt{y^2 - 2018}) = 2018.$$

Find the value of

$$2019xy - 2018y^2 + 2017x^2 - 2017^2.$$

Problem 5.26 (ZIML Varsity Jan 2019) Let x be a real number, find the maximum value of

$$\sqrt{4x - 3} + \sqrt{2 - x},$$

rounded to the nearest tenth if necessary.

Problem 5.27 (ZIML Varsity Feb 2018) Find the smallest M such that the inequality

$$2ab + 3bc + 2cd \leq M(a^2 + b^2 + c^2 + d^2)$$

holds for all groups of real numbers a, b, c, d.

Problem 5.28 (ZIML Varsity Feb 2019) Let $a_1 = \dfrac{1}{10}$ and $a_{k+1} = a_k^2 + a_k (k = 1, 2, 3, \ldots)$. Define

$$S = \frac{1}{a_1 + 1} + \frac{1}{a_2 + 1} + \cdots + \frac{1}{a_{90} + 1},$$

find the value of $\lfloor S \rfloor$ (which is the greatest integer not exceeding S).

Problem 5.29 (ZIML Varsity January 2019) Let a, b, c be real numbers satisfying

$$\begin{cases} a^2 + b^2 + c^2 = 1, \\ a\left(\dfrac{1}{b} + \dfrac{1}{c}\right) + b\left(\dfrac{1}{c} + \dfrac{1}{a}\right) + c\left(\dfrac{1}{a} + \dfrac{1}{b}\right) = -3. \end{cases}$$

Find the minimum possible value for $a + b + c$.

Problem 5.30 (ZIML Varsity January 2019) Let a, b, and c be positive real numbers satisfying the following system of equations:

$$\begin{cases} 8\sqrt{a-b+c} \cdot \sqrt{a+b-c} = a\sqrt{bc}, \\ 10\sqrt{b-c+a} \cdot \sqrt{b+c-a} = b\sqrt{ca}, \\ 12\sqrt{c-a+b} \cdot \sqrt{c+a-b} = c\sqrt{ab}. \end{cases}$$

Find the values of a, b, and c.

Hint: Geometric interpretation: a, b, and c can be seen as the three sides of a triangle.

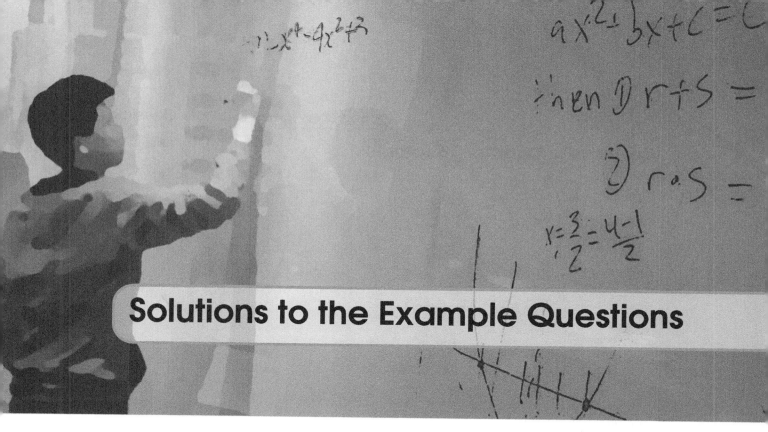

Solutions to the Example Questions

In the sections below you will find solutions to all of the Example Questions contained in this book.

Practice questions are meant to be used for homework, so their answers and solutions are not included. Teachers or math coaches may contact Areteem at info@areteem.org for answer keys and options for purchasing a Teachers' Edition of the course.

1   Solutions to Chapter 1 Examples

Problem 1.1 Show that

$$x^{\log_a y} = y^{\log_a x}$$

Solution

Take logarithm with base a on both sides:

$$\log_a x^{\log_a y} = \log_a y \cdot \log_a x$$

and

$$\log_a y^{\log_a x} = \log_a x \cdot \log_a y,$$

and clearly both sides are the same.

Remark

The identity in Problem 1.1 is a very nice one. We will use it later on in other examples.

Problem 1.2 Assume function $f(x) = \log_2(x^2 + ax + 1)$ is well-defined on all $x \in \mathbb{R}$. Find the range of possible values of a.

Answer

$-2 < a < 2$

Solution

Since $x^2 + ax + 1 > 0$ for all real numbers x, the equation $x^2 + ax + 1 = 0$ has no real roots. Therefore the discriminant $a^2 - 4 < 0$, and thus $-2 < a < 2$.

Problem 1.3 Let $x > 0$ be a real number. Given that $\log_{\sqrt{2}} x = 100$, compute $\log_{\sqrt{x}} 2$.

Answer

$\dfrac{1}{25}$

Solution

Convert to exponential form: $x = (\sqrt{2})^{100} = 2^{50}$, so $\log_{\sqrt{x}} 2 = \dfrac{1}{\log_2 \sqrt{x}} = \dfrac{1}{25}$.

Problem 1.4 Solve the equation for x:

$$(\log_5 x)^2 = \log_5 x^2.$$

Answer

$x = 1$ or $x = 25$

Solution

We get

$$(\log_5 x)^2 = 2 \log_5 x;$$

let $y = \log_5 x$, then $y^2 = 2y$, so $y = 0$ or $y = 2$. If $y = 0$, then $x = 1$; if $y = 2$, then $x = 25$. Verifying, both are solutions.

Problem 1.5 Let $a > 0$ and $a \neq 1$, Show that $\log_{a^2} X = \dfrac{3}{2} \log_{a^3} X$ for all $X > 0$.

Solution

Change to base a:

$$\log_{a^2} X = \frac{\log_a X}{\log_a a^2} = \frac{\log_a X}{2},$$

$$\log_{a^3} X = \frac{\log_a X}{\log_a a^3} = \frac{\log_a X}{3},$$

therefore $\log_{a^2} X = \dfrac{3}{2} \log_{a^3} X$.

Problem 1.6 Show that $\log_{10} 2$ is irrational.

Solution

We use proof by contradiction. First note that

$$0 = \log_{10} 1 < \log_{10} 2 < \log_{10} 10 = 1.$$

Assume $\log_{10} 2$ is a rational number, i.e. $\log_{10} 2 = \dfrac{p}{q}$ where p and q are positive integers with $\gcd(p,q) = 1$, and $p < q$. Then

$$
\begin{aligned}
10^{p/q} &= 2, \\
10^{p} &= 2^{q}, \\
5^{p} \cdot 2^{p} &= 2^{q}, \\
5^{p} &= 2^{q-p}.
\end{aligned}
$$

Since $p < q$, both sides are integers, but the left hand side is odd and the right hand side is even, a contradiction.

Therefore $\log_{10} 2$ is irrational.

Problem 1.7 Evaluate

$$
\frac{5}{\log_2 2016^3} + \frac{2}{\log_3 2016^3} + \frac{1}{\log_7 2016^3}.
$$

Answer

$\dfrac{1}{3}$

Solution

Change the base to 2016^3:

$$
\begin{aligned}
\frac{5}{\log_2 2016^3} + \frac{2}{\log_3 2016^3} + \frac{1}{\log_7 2016^3} &= 5\log_{2016^3} 2 + 2\log_{2016^3} 3 + \log_{2016^3} 7 \\
&= \log_{2016^3}(2^5 \times 3^2 \times 7) \\
&= \log_{2016^3} 2016 \\
&= \frac{1}{3}.
\end{aligned}
$$

Problem 1.8 Given that $1 < a < b < a^2$, arrange the following four numbers in increasing order:

$$
2, \quad \log_a b, \quad \log_b a, \quad \log_{ab} a^2.
$$

Answer

$\log_b a, \log_{ab} a^2, \log_a b, 2$

Solution

Since $b > a > 1$, the functions $\log_a x$ and $\log_b x$ are both increasing functions. So

$$1 = \log_a a < \log_a b < \log_a a^2 = 2.$$

Clearly,

$$\log_b a = \frac{1}{\log_a b} < 1,$$

and

$$\log_{ab} a^2 < \log_{ab} ab = 1.$$

Now it remains to compare $\log_b a$ and $\log_{ab} a^2$. We compare their reciprocals, using the fact that $\log_a b > 1$:

$$\frac{1}{\log_{ab} a^2} = \frac{1}{2\log_{ab} a} = \frac{1}{2}\log_a ab = \frac{1}{2}(\log_a a + \log_a b) = \frac{1}{2}(1 + \log_a b) < \log_a b = \frac{1}{\log_b a}.$$

So $\log_{ab} a^2 > \log_b a$. Therefore the four numbers in increasing order:

$$\log_b a, \log_{ab} a^2, \log_a b, 2.$$

Problem 1.9 Evaluate $\lg\left(\sqrt{3 + \sqrt{5}} + \sqrt{3 - \sqrt{5}}\right)$. Here lg means logarithm with base 10.

Answer

$\dfrac{1}{2}$

Solution

Simplify the argument first: let $x = \sqrt{3 + \sqrt{5}} + \sqrt{3 - \sqrt{5}}$, then

$$\begin{aligned}
x^2 &= \left(\sqrt{3 + \sqrt{5}} + \sqrt{3 - \sqrt{5}}\right)^2 \\
&= 3 + \sqrt{5} + 2\sqrt{3 + \sqrt{5}} \cdot \sqrt{3 - \sqrt{5}} + 3 - \sqrt{5} \\
&= 6 + 2\sqrt{9 - 5} \\
&= 10,
\end{aligned}$$

so $x = \sqrt{10}$, and $\lg x = \dfrac{1}{2}$.

Problem 1.10 Evaluate $6^{\lg 40} \times 5^{\lg 36}$. Here lg represents logarithm with base 10.

Answer

216

Solution

Using the identity $x^{\log_a y} = y^{\log_a x}$:

$$6^{\lg 40} \times 5^{\log 36} = 40^{\lg 6} \times 5^{2\lg 6} = (40 \times 25)^{\lg 6} = 1000^{\lg 6} = 6^{\lg 1000} = 6^3 = 216.$$

Problem 1.11 The real numbers $x, y,$ and z are all greater than 1, and w is a positive number such that $\log_x w = 45$, $\log_y w = 60$, and $\log_{xyz} w = 15$. Find $\log_z w$.

Answer

36

Solution

Change all the bases to w:

$$\log_w x = \frac{1}{45}, \quad \log_w y = \frac{1}{60}, \quad \log_w xyz = \frac{1}{15}.$$

Therefore,

$$\log_w z = \frac{1}{15} - \frac{1}{45} - \frac{1}{60} = \frac{1}{36},$$

and then

$$\log_z w = 36.$$

Problem 1.12 Let x_1 be a root of the equation $\log_3 x + x - 3 = 0$, and x_2 be a root of the equation $3^x + x - 3 = 0$, find the value of $x_1 + x_2$.

Answer

3

Solution

We know that

$$\begin{aligned} \log_3 x_1 + x_1 &= 3 \\ 3^{x_2} + x_2 &= 3. \end{aligned}$$

The second equation above can be written as

$$\log_3(3^{x_2}) + 3^{x_2} = 3.$$

For $x > 0$, $f(x) = \log_3 x + x$ is an increasing function, and since $f(x_1) = f(3^{x_2}) = 3$, we have $x_1 = 3^{x_2}$. Therefore $x_1 + x_2 = 3^{x_2} + x_2 = 3$.

Problem 1.13 Given that a, b, c are all positive and not equal to 1, simplify the following:

$$a^{\log_2(b/c)} \cdot b^{\log_2(c/a)} \cdot c^{\log_2(a/b)}$$

Answer

1

Solution

Let

$$x = a^{\log_2(b/c)} \cdot b^{\log_2(c/a)} \cdot c^{\log_2(a/b)},$$

and take logarithm with base 2,

$$\begin{aligned} \log_2 x &= \log_2 \frac{b}{c} \cdot \log_2 a + \log_2 \frac{c}{a} \cdot \log_2 b + \log_2 \frac{a}{b} \cdot \log_2 c \\ &= (\log_2 b - \log_2 c)\log_2 a + (\log_2 c - \log_2 a)\log_2 b + (\log_2 a - \log_2 b)\log_2 c \\ &= 0, \end{aligned}$$

therefore $x = 1$.

Problem 1.14 Let x and y be positive and $\log_9 x = \log_{12} y = \log_{16}(x+y)$, find $\frac{y}{x}$.

Answer

$$\frac{1 + \sqrt{5}}{2}$$

Solution

Let $\log_9 x = \log_{12} y = \log_{16}(x+y) = k$, then $x = 9^k, y = 12^k, x+y = 16^k$. Thus $y^2 = x(x+y)$, which is

$$y^2 - xy - x^2 = 0.$$

Let $z = \dfrac{y}{x}$,

$$z^2 - z - 1 = 0$$

so $z = \dfrac{1 \pm \sqrt{5}}{2}$. Since $x > 0, y > 0$, we get $z > 0$, so $z = \dfrac{1 + \sqrt{5}}{2}$.

Problem 1.15 Let m and n be positive integers, $a > 0$ and $a \neq 1$, and

$$\log_a m + \log_a \left(1 + \frac{1}{m}\right) + \log_a \left(1 + \frac{1}{m+1}\right) + \cdots + \log_a \left(1 + \frac{1}{m+n-1}\right)$$
$$= \log_a m + \log_a n,$$

find the values of m and n.

Answer

$m = n = 2$

Solution

First simplify the left hand side:

$$
\begin{aligned}
& \log_a m + \log_a \left(1 + \frac{1}{m}\right) + \log_a \left(1 + \frac{1}{m+1}\right) + \cdots + \log_a \left(1 + \frac{1}{m+n-1}\right) \\
= \ & \log_a m + \log_a \left(\frac{m+1}{m}\right) + \log_a \left(\frac{m+2}{m+1}\right) + \cdots + \log_a \left(\frac{m+n}{m+n-1}\right) \\
= \ & \log_a m \cdot \frac{m+1}{m} \cdot \frac{m+2}{m+1} \cdots \frac{m+n}{m+n-1} \\
= \ & \log_a (m+n).
\end{aligned}
$$

Thus

$$\log_a (m+n) = \log_a mn,$$

then

$$m + n = mn.$$

Since m and n are positive integers, we use the "completing the rectangle" technique,

$$
\begin{aligned}
mn - m - n &= 0, \\
mn - m - n + 1 &= 1, \\
(m-1)(n-1) &= 1.
\end{aligned}
$$

Here the only possibility is $m - 1 = n - 1 = 1$, so $m = n = 2$.

2   Solutions to Chapter 2 Examples

Problem 2.1 Convert between the rectangular, polar, and exponential forms: write each of these numbers in the other forms.

(a) i

Answer

$$i = \cos\frac{\pi}{2} + i\sin\frac{\pi}{2} = e^{i\pi/2}$$

(b) $1 + i$

Answer

$$1 + i = \sqrt{2}\left(\cos\frac{\pi}{4} + i\sin\frac{\pi}{4}\right) = \sqrt{2}e^{i\pi/4}$$

(c) $4\left(\cos\frac{\pi}{6} + i\sin\frac{\pi}{6}\right)$

Answer

$$4\left(\cos\frac{\pi}{6} + i\sin\frac{\pi}{6}\right) = 4e^{i\pi/6} = 2\sqrt{3} + 2i$$

Problem 2.2 Convert complex numbers among different forms:

(a) Let a and b be real numbers, convert $a + bi$ to polar form.

Solution

If $a \neq 0$: $a + bi = \sqrt{a^2 + b^2}(\cos\theta + i\sin\theta) = \sqrt{a^2 + b^2}e^{i\theta}$ where $\theta = \arctan\frac{b}{a}$ (if $a + bi$ is in Quadrant I or IV) or $\arctan\frac{b}{a} + \pi$ (otherwise).

If $a = 0, b \geq 0$: $bi = b\left(\cos\frac{\pi}{2} + i\sin\frac{\pi}{2}\right)$;

If $a = 0, b < 0$: $bi = |b|\left(\cos\frac{3\pi}{2} + i\sin\frac{3\pi}{2}\right)$.

(b) Convert $re^{i\theta}$ to rectangular form.

Answer

$r\cos\theta + ir\sin\theta$

Solution

$re^{i\theta} = r(\cos\theta + i\sin\theta) = r\cos\theta + ir\sin\theta.$

Problem 2.3 What are the sets of points satisfying the following? Draw the diagrams.

(a) $|z| \leq 2$

Answer

Circular region centered at the Origin with radius 2, boundary included

(b) $\Re z > \dfrac{1}{2}$ (The real part of z is greater than $\dfrac{1}{2}$)

Answer

The half plane whose x-coordinate is greater than $\dfrac{1}{2}$

(c) $\Re z = \Im z$

Answer

The line $y = x$

Problem 2.4 $|z_1 + z_2|^2 + |z_1 - z_2|^2 = 2\left(|z_1|^2 + |z_2|^2\right)$. What is the geometrical interpretation of this identity?

Solution

$$
\begin{aligned}
&|z_1 + z_2|^2 + |z_1 - z_2|^2 \\
= {}& (z_1 + z_2) \cdot \overline{(z_1 + z_2)} + (z_1 - z_2) \cdot \overline{(z_1 - z_2)} \\
= {}& z_1 \cdot \overline{z_1} + z_1 \cdot \overline{z_2} + z_2 \cdot \overline{z_1} + z_2 \cdot \overline{z_2} + z_1 \cdot \overline{z_1} - z_1 \cdot \overline{z_2} - z_2 \cdot \overline{z_1} + z_2 \cdot \overline{z_2} \\
= {}& 2\left(|z_1|^2 + |z_2|^2\right).
\end{aligned}
$$

The geometrical interpretation: In a parallelogram, the sum of the squares of the diagonals equals the sum of squares of all four sides.

Problem 2.5 If $|a| = |b| = 1$ and $a + b + 1 = 0$, what are a and b?

Answer

$$-\frac{1}{2} \pm \frac{\sqrt{3}}{2}i$$

Solution

Let $a = \cos\alpha + i\sin\alpha$ and $b = \cos\beta + i\sin\beta$, then $\cos\alpha + i\sin\alpha + \cos\beta + i\sin\beta + 1 = 0$. Here we get $\cos\alpha + \cos\beta + 1 = 0$ and $\sin\alpha + \sin\beta = 0$. Hence, (1) $\alpha = 2k\pi - \beta$ or (2) $\alpha = (2k+1)\pi + \beta$. In case (1), $\cos\alpha = \cos\beta$, thus they are both $-\frac{1}{2}$, and then we have $a = -\frac{1}{2} \pm \frac{\sqrt{3}}{2}i$, and $b = -1 - a = -\frac{1}{2} \mp \frac{\sqrt{3}}{2}i$. In case (2), $\cos\alpha = -\cos\beta$ and there is no solution.

Conclusion: a and b are $-\frac{1}{2} \pm \frac{\sqrt{3}}{2}i$.

Problem 2.6 Find the square roots of i.

Answer

$$\pm\left(\frac{\sqrt{2}}{2} + \frac{\sqrt{2}}{2}i\right).$$

Solution

$i = e^{i\pi/2}$, so the square root of i is $\pm e^{i\pi/4} = \pm\left(\cos\frac{\pi}{4} + i\sin\frac{\pi}{4}\right) = \pm\left(\frac{\sqrt{2}}{2} + \frac{\sqrt{2}}{2}i\right)$.

Problem 2.7 If $f(z) = z^{3m+1} + z^{3m+2} + 1$. Show that $f(z)$ is divisible by $z^2 + z + 1$.

Solution

Let ω_1 and ω_2 represent the cube roots of unity. Then ω_1 and ω_2 are the roots of $z^2 + z + 1$. We also know that $\omega_1^3 = \omega_2^3 = 1$, thus $\omega_1^{3m+1} = \omega_1$, and $\omega_2^{3m+2} = \omega_1^2$. Thus $f(\omega_1) = \omega_1^{3m+1} + \omega_1^{3m+2} + 1 = \omega_1 + \omega_2^2 + 1 = 0$. By Factor Theorem, $f(z)$ has a factor $(z - \omega_1)$. Similarly, the same is true for ω_2. Thus $(z - \omega_1)(z - \omega_2) = z^2 + z + 1$ must be a factor of $f(z)$.

Problem 2.8 Let a and b be real numbers. Given that $2 + ai$ and $b + i$ are the two roots of the quadratic equation $x^2 + px + q = 0$ where p and q are real numbers. What are p and q?

Answer

$p = -4, q = 5$

Solution

By Vieta's formulas, $p = -(2 + ai + b + i) = -(b + 2) - (a + 1)i$, and $q = (2 + ai)(b + i) = (2b - a) + (2 + ab)i$. Since p and q are real numbers, $a + 1 = 0$ and $2 + ab = 0$. Thus $a = -1$, $b = 2$. Therefore $p = -4, q = 5$.

Problem 2.9 Find the remainder when $x^{1001} - 1$ is divided by $x^4 + x^3 + 2x^2 + x + 1$.

Answer

$-x^3 + x^2$

Solution

Let $f(x) = x^{1001} - 1$, $d(x) = x^4 + x^3 + 2x^2 + x + 1 = (x^2 + 1)(x^2 + x + 1)$, then we have the division formula

$$f(x) = d(x) \cdot q(x) + r(x),$$

where $d(x)$ is a divisor and $r(x)$ is the remainder, and the degree of polynomial $r(x)$ should be lower than that of $d(x)$. So $\deg r \leq 3$.

Let $r(x) = ax^3 + bx^2 + cx + d$, and let $\omega = -\dfrac{1}{2} + \dfrac{\sqrt{3}}{2}i$ be the cube root of unity. Then we have $\omega^3 = 1$ and $\omega^2 = -1 - \omega$. Since $d(x) = (x^2 + 1)(x^2 + x + 1)$, $d(\pm i) = 0$ and $d(\omega) = 0$ and $d(\omega^2) = 0$.

Let $x = i$, so $f(i) = r(i)$, that is,

$$i - 1 = -ai - b + ci + d = (d - b) + (c - a)i;$$

Let $x = \omega$, then $f(\omega) = r(\omega)$, so

$$-2 - \omega = a + (-b - b\omega) + c\omega + d,$$

$$-2 + \frac{1}{2} - \frac{\sqrt{3}}{2}i = \left(a - \frac{1}{2}b - \frac{1}{2}c + d \right) - \frac{\sqrt{3}}{2}(b - c)i.$$

Comparing the real parts and imaginary parts of the two equations,

$$
\begin{aligned}
d - b &= -1, \\
c - a &= 1, \\
a + d - \frac{1}{2}(b+c) &= -\frac{3}{2}, \\
b - c &= 1.
\end{aligned}
$$

Solve and get $a = -1, b = 1, c = d = 0$, so $r(x) = -x^3 + x^2$.

Problem 2.10 Factor the polynomial: $x^8 + x^6 + x^4 + x^2 + 1$.

Answer

$(x^4 - x^3 + x^2 - x + 1)(x^4 + x^3 + x^2 + x + 1)$

Solution

Notice that $x^{10} - 1 = (x^2 - 1)(x^8 + x^6 + x^4 + x^2 + 1)$, the zeros of the given polynomial are part of the 10th roots of unity.

$$
\begin{aligned}
x^8 + x^6 + x^4 + x^2 + 1 &= \frac{x^{10} - 1}{x^2 - 1} \\
&= \frac{(x^5 + 1)(x^5 - 1)}{(x+1)(x-1)} \\
&= \frac{x^5 + 1}{x+1} \cdot \frac{x^5 - 1}{x - 1} \\
&= (x^4 - x^3 + x^2 - x + 1)(x^4 + x^3 + x^2 + x + 1).
\end{aligned}
$$

Problem 2.11 Use complex numbers to prove the following theorem:
In quadrilateral $ABCD$, let E, F, G, H be the midpoints of $\overline{AB}, \overline{BC}, \overline{CD}, \overline{DA}$ respectively, and let M and N be the midpoints of $\overline{AC}$ and $\overline{BD}$ respectively, then $\overline{EG}, \overline{FH}$ and $\overline{MN}$ are concurrent.

Solution

Let complex numbers $a, b, c, d, e, f, g, h, m, n$ represent the points $A, B, C, D, E, F, G,$

H, M, N respectively. Then

$$e = \frac{a+b}{2}, \qquad f = \frac{b+c}{2},$$
$$g = \frac{c+d}{2}, \qquad h = \frac{d+a}{2},$$
$$m = \frac{a+c}{2}, \qquad n = \frac{b+d}{2}.$$

Now we find the mid points of $\overline{EG}, \overline{FH}$, and $\overline{MN}$, and they all turn out to be $\frac{a+b+c+d}{4}$. This means $\overline{EG}, \overline{FH}$, and $\overline{MN}$ are concurrent.

Problem 2.12 Let $\varepsilon = \cos\frac{2\pi}{n} + i\sin\frac{2\pi}{n}$, evaluate

$$(1-\varepsilon)(1-\varepsilon^2)\cdots(1-\varepsilon^{n-1})$$

Answer

n

Solution

Since $\varepsilon, \varepsilon^2, \ldots, \varepsilon^{n-1}$ are all n^{th} roots of unity,

$$(x-\varepsilon)(x-\varepsilon^2)\cdots(x-\varepsilon^{n-1}) = x^{n-1}+x^{n-2}+\cdots+x+1.$$

Let $x = 1$, then we have

$$(1-\varepsilon)(1-\varepsilon^2)\cdots(1-\varepsilon^{n-1}) = n.$$

Problem 2.13 On the complex plane, given points $B(1), C(2+i)$, ABC is an equilateral triangle. Find the possible positions for A.

Answer

$\frac{3\pm\sqrt{3}}{2} + \frac{1\mp\sqrt{3}}{2}i$

Solution

Let $b = 1$ and $c = 2 + i$, and a be the complex number representing point A. Since $\triangle ABC$ is equilateral, side $\overline{AB}$ is side $\overline{BC}$ rotated $60°$ in either direction. Thus

$$a - b = (c - b)\operatorname{cis}(\pm 60°).$$

So either

$$a = 1 + (1 + i)\left(\frac{1}{2} + \frac{\sqrt{3}}{2}i\right) = \frac{3 - \sqrt{3}}{2} + \frac{1 + \sqrt{3}}{2}i,$$

or

$$a = 1 + (1 + i)\left(\frac{1}{2} - \frac{\sqrt{3}}{2}i\right) = \frac{3 + \sqrt{3}}{2} + \frac{1 - \sqrt{3}}{2}i.$$

Problem 2.14 Evaluate: $1 + \binom{n}{3} + \binom{n}{6} + \cdots + \binom{n}{3m}$ where $3m$ is the maximum multiple of 3 not exceeding n.

Answer

$\dfrac{1}{3}\left(2^n + 2\cos\dfrac{n\pi}{3}\right)$

Solution

By Binomial Theorem,

$$(1 + x)^n = \binom{n}{0} + \binom{n}{1}x + \binom{n}{2}x^2 + \cdots + \binom{n}{n-1}x^{n-1} + \binom{n}{n}x^n.$$

Let $\omega = \operatorname{cis}\dfrac{2\pi}{3}$ be the imaginary cube root of unity, then $\omega^3 = 1$ and $1 + \omega + \omega^2 = 0$, so

$$1 + \omega^k + \omega^{2k} = \begin{cases} 3, & \text{if } 3 \mid k; \\ 0, & \text{if } 3 \nmid k. \end{cases}$$

Let $x = 1, \omega, \omega^2$, and add the three equations,

$$2^n + (1 + \omega)^n + (1 + \omega^2)^n = 3\left(1 + \binom{n}{3} + \binom{n}{6} + \cdots + \binom{n}{3m}\right).$$

We know that $1 + \omega = -\omega^2 = -\operatorname{cis}\dfrac{4\pi}{3} = \operatorname{cis}\dfrac{\pi}{3}$, and $1 + \omega^2 = -\omega = \operatorname{cis}\left(-\dfrac{\pi}{3}\right)$,

$$2^n + (1 + \omega)^n + (1 + \omega^2)^n = 2^n + \operatorname{cis}\frac{n\pi}{3} + \operatorname{cis}\left(-\frac{n\pi}{3}\right) = 2^n + 2\cos\frac{n\pi}{3},$$

therefore

$$1 + \binom{n}{3} + \binom{n}{6} + \cdots + \binom{n}{3m} = \frac{1}{3}\left(2^n + 2\cos\frac{n\pi}{3}\right).$$

Problem 2.15 Simplify the sums: $\displaystyle\sum_{k=0}^{n-1}\cos k\theta$ and $\displaystyle\sum_{k=0}^{n-1}\sin k\theta$

Answer

$\dfrac{1 - \cos\theta - \cos n\theta + \cos(n-1)\theta}{2 - 2\cos\theta}$ and $\dfrac{\sin\theta - \sin n\theta + \sin(n-1)}{2 - 2\cos\theta}$ (the answers can be written in different but equivalent forms)

Solution

Let $C = \displaystyle\sum_{k=0}^{n-1}\cos k\theta$ and $S = \displaystyle\sum_{k=0}^{n-1}\sin k\theta$, then

$$\begin{aligned}
C + iS &= \sum_{k=0}^{n-1}(\cos k\theta + i\sin k\theta)\\[2mm]
&= \sum_{k=0}^{n-1}e^{ik\theta}\\[2mm]
&= \frac{1 - e^{in\theta}}{1 - e^{i\theta}}\\[2mm]
&= \frac{1 - \cos n\theta - i\sin n\theta}{1 - \cos\theta - i\sin\theta}\\[2mm]
&= \frac{(1 - \cos\theta - \cos n\theta + \cos(n-1)\theta) + i(\sin\theta - \sin n\theta + \sin(n-1)\theta)}{2 - 2\cos\theta}.
\end{aligned}$$

Comparing the real parts and imaginary parts,

$$C = \frac{1 - \cos\theta - \cos n\theta + \cos(n-1)\theta}{2 - 2\cos\theta},$$

and

$$S = \frac{\sin\theta - \sin n\theta + \sin(n-1)}{2 - 2\cos\theta}.$$

The results can be written in different forms, after trigonometric transformations. The following is another possible form:

$$C = \frac{\sin\dfrac{n\theta}{2}\cos\dfrac{(n-1)\theta}{2}}{\sin\dfrac{\theta}{2}}, \quad S = \frac{\sin\dfrac{n\theta}{2}\sin\dfrac{(n-1)\theta}{2}}{\sin\dfrac{\theta}{2}}.$$

3   Solutions to Chapter 3 Examples

Problem 3.1 Let $x = 100$ be a root of quadratic equation $ax^2 + bx + a = 0$. What is the other root?

Answer

1/100

Solution

Let y be the other root. By Vieta's theorem, then

$$xy = \frac{a}{a} = 1.$$

Since $x = 100$, we get $y = \dfrac{1}{100}$.

Problem 3.2 The equation in x: $x^2 + px + q = 0$ has two nonzero integer roots, and $p + q = 198$. What is p?

Answer

-202.

Solution

Let x_1 and x_2 be the roots. According to Vieta Theorem, $x_1 + x_2 = -p$ and $x_1 x_2 = q$. So we have

$$x_1 x_2 - x_1 - x_2 = 198.$$

Add 1 to both sides (this method is called "completing the rectangle"),

$$x_1 x_2 - x_1 - x_2 + 1 = 199,$$

and

$$(x_1 - 1)(x_2 - 1) = 199.$$

Since 199 is a prime number, and the roots are integers, the only possibilities are $x_1 - 1 = 1, x_2 - 1 = 199$ and $x_1 - 1 = -1, x_2 - 1 = -199$. We are also given that the roots are nonzero, so $x_1 = 2$ and $x_2 = 200$. So $p = -202$.

Problem 3.3 Find real solutions for

$$\begin{cases} x+y & =6 \\ xy-z^2 & =9 \end{cases}$$

Answer

$x=y=3, z=0$

Solution

Rewrite the second equation as $xy=z^2+9$, then by Vieta's formulas, x and y are real roots for $t^2-6t+(z^2+9)=0$, whose discriminant is $-4z^2 \geq 0$, so $z=0$, and $x=y=3$.

Problem 3.4 $(x^2+x+1)(x^2+x+2)=12$

Answer

$-2, 1$

Solution

Let $y=x^2+x$. This gives $(y+1)(y+2)=12$ so $y^2+3y-10=(y+5)(y-2)=0$, and we get $y=-5, y=2$.
Then plug back in x: $x^2+x=-5$ so $x^2+x+5=0$ (no real solutions) or $x^2+x=2$ so $x^2+x-2=0$. Therefore the solutions are $x=-2, x=1$.

Problem 3.5 $(x^2+x-2)^2+(2x^2-5x+3)^2=(3x^2-4x+1)^2$

Answer

$-2, 1, 3/2$ (1 is a double root)

Solution

Let $A=x^2+x-2, B=2x^2-5x+3$, then $A+B=3x^2-4x+1$, therefore the equation is $A^2+B^2=(A+B)^2$, so $AB=0$. Factoring,

$$(x-1)(x+2)(x-1)(2x-3)=0,$$

so the roots are $-2, 1, 3/2$ (1 is double root).

Problem 3.6 Find the real solutions: $x^2 + x + 1 = \dfrac{2}{x^2 + x}$

Answer

$$x = \frac{-1 \pm \sqrt{5}}{2}$$

Solution

Let $y = x^2 + x$, so $x = \dfrac{-1 \pm \sqrt{5}}{2}$.

Problem 3.7 Find the real roots: $x^4 + (x-2)^4 = 706$.

Answer

$5, -3$

Solution

Let $y = x - 1$, then the equation becomes

$$(y+1)^4 + (y-1)^4 = 706.$$

Expanding the left hand side and simplify

$$
\begin{aligned}
2y^4 + 12y^2 + 2 &= 706, \\
y^4 + 6y^2 - 352 &= 0, \\
(y^2 - 16)(y^2 + 22) &= 0, \\
(y+4)(y-4)(y^2 + 22) &= 0.
\end{aligned}
$$

This has real roots $y = 4$ and $y = -4$, so the real roots of the original equation are $x = 5$ and $x = -3$.

Problem 3.8 Find the real roots: $2x^4 - 9x^3 + 14x^2 - 9x + 2 = 0$.

Answer

$1/2, 1, 2$ (1 is a double root)

Solution

Clearly $x = 0$ is not a root. Divide both sides by x^2, we get

$$2x^2 - 9x + 14 - \frac{9}{x} + \frac{2}{x^2} = 0,$$

rearranging the terms,

$$2\left(x^2+\frac{1}{x^2}\right)-9\left(x+\frac{1}{x}\right)+14=0.$$

Let $y=x+\frac{1}{x}$, then $y^2=x^2+\frac{1}{x^2}+2$, so the equation becomes

$$\begin{aligned}2(y^2-2)-9y+14 &= 0,\\ 2y^2-9y+10 &= 0,\\ (y-2)(2y-5) &= 0,\end{aligned}$$

which gives two solutions: $y=2$ and $y=\frac{5}{2}$.

If $y=2$, $x+\frac{1}{x}=2$, $x^2-2x+1=0$, so $x=1$ is a double root.

If $y=\frac{5}{2}$, $x+\frac{1}{x}=\frac{5}{2}$, so $2x^2-5x+2=0$, $(x-2)(2x-1)=0$, thus the roots are $x=2$ and $x=\frac{1}{2}$.

Therefore, the roots of the original equation are $x=2, x=\frac{1}{2}, x=1$, where $x=1$ is a double root.

Problem 3.9 Let x,y be **positive integers**, and satisfy

$$\begin{cases} xy+x+y &= 71 \\ x^2y+xy^2 &= 880 \end{cases}$$

Find the value of x^2+y^2.

Answer

146

Solution

Let $u=x+y$, $v=xy$, then $u+v=71$ and $uv=880$. By Vieta's formulas, u and v are roots of the quadratic equation $t^2-71t+880=0$. This equation can be transformed to $(t-16)(t-55)=0$, thus $u=16, v=55$ or $u=55, v=16$.

If $x+y=16, xy=55$, by Vieta's formulas, x and y are roots of the equation $z^2-16z+55=0$, so we get roots 5 and 11, which are integers, thus $x^2+y^2=(x+y)^2-2xy=16^2-2\times55=146$.

If $x+y=55, xy=1$, by Vieta's formulas, x and y are roots of the equation $z^2-55z+16=0$, which has no integer roots.

In conclusion, the answer is 146.

Problem 3.10 Find the real solutions:

$$\begin{cases} x^3 + y^3 + x^3 y^3 &= 17 \\ x + y + xy &= 5 \end{cases}$$

Answer

$(1, 2)$ and $(2, 1)$

Solution

Let $u = x + y, v = xy$, then $u + v = 5, u^3 + v^3 - 3uv = 17$. Also, $u^3 + v^3 - 3uv = (u + v)^3 - 3uv(u + v) - 3uv = 17$. Using the fact that $u + v = 5$, we get $125 - 18uv = 17$, thus $uv = 6$. So u and v are roots of the equation $t^2 - 5t + 6 = 0$, and thus $(u, v) = (3, 2)$ or $(u, v) = (2, 3)$.
If $(u, v) = (3, 2)$, $x + y = 3, xy = 2$, then $(x, y) = (1, 2)$ or $(2, 1)$.
If $(u, v) = (2, 3)$, $x + y = 2, xy = 3$, there are no real solutions.
Therefore, $(1, 2)$ and $(2, 1)$ are the only real solutions.

Problem 3.11 Find the real solutions: $5x^2 + x - x\sqrt{5x^2 - 1} - 2 = 0$

Answer

$\pm\dfrac{\sqrt{10}}{5}$

Solution

Let $y = \sqrt{5x^2 - 1}$. Roots: $\pm\dfrac{\sqrt{10}}{5}$ and $1/2, -1$, but $1/2$ and -1 are extraneous.

Problem 3.12 Find the real solutions: $\dfrac{x - 1}{x + 1} + \dfrac{x - 4}{x + 4} = \dfrac{x - 2}{x + 2} + \dfrac{x - 3}{x + 3}$

Answer

0 and $-5/2$

Solution

Add 1 to each fraction, then

$$\frac{2x}{x + 1} + \frac{2x}{x + 4} = \frac{2x}{x + 2} + \frac{2x}{x + 3},$$

clearly $x = 0$ is a solution.

For the case $x \neq 0$, canceling x from both sides,

$$\frac{1}{x+1} + \frac{1}{x+4} = \frac{1}{x+2} + \frac{1}{x+3},$$

thus

$$\frac{2x+5}{(x+1)(x+4)} = \frac{2x+5}{(x+2)(x+3)},$$

here we can see that $x = -5/2$ is also a root.

Again, consider the case where $x \neq 0$ and $x \neq -5/2$, then

$$(x+1)(x+4) = (x+2)(x+3),$$

so

$$x^2 + 5x + 4 = x^2 + 5x + 6,$$

which has no solution. Therefore the solutions are $x = 0$ and $x = -5/2$.

Problem 3.13 Solve for x: $\left(\sqrt[5]{3}\right)^x + \left(\sqrt[10]{3}\right)^{x-10} = 84$

Answer

20

Solution

Problem 3.14 Find the real roots: $\sqrt{5 - \sqrt{5-x}} = x$.

Answer

$$\frac{\sqrt{21} - 1}{2}$$

Solution

First of all, $x \geq 0$. Square both sides,

$$5 - \sqrt{5-x} = x^2,$$

so

$$5 - x^2 = \sqrt{5-x}.$$

Note that $5 - x^2 \geq 0$. Thus $0 \leq x \leq \sqrt{5}$.

Squaring again,

$$5^2 - 2 \cdot 5 \cdot x^2 + x^4 = 5 - x.$$

We treat the number "5" as the unknown and x as a constant, then the equation is a quadratic equation in "5":

$$5^2 - (2x^2 + 1) \cdot 5 + (x^4 + x) = 0.$$

Applying the quadratic formula,

$$\begin{aligned}
5 &= \frac{(2x^2 + 1) \pm \sqrt{(2x^2 + 1)^2 - 4(x^4 + x)}}{2} \\[2mm]
&= \frac{(2x^2 + 1) \pm \sqrt{4x^4 + 4x^2 + 1 - 4x^4 - 4x}}{2} \\[2mm]
&= \frac{(2x^2 + 1) \pm \sqrt{4x^2 - 4x + 1}}{2} \\[2mm]
&= \frac{(2x^2 + 1) \pm (2x - 1)}{2}
\end{aligned}$$

So $5 = x^2 + x$ or $5 = x^2 - x + 1$.

If $5 = x^2 + x$, $x^2 + x - 5 = 0$, then $x = \dfrac{-1 \pm \sqrt{21}}{2}$;

If $5 = x^2 - x + 1$, $x^2 - x - 4 = 0$, $x = \dfrac{1 \pm \sqrt{17}}{2}$.

Since $0 \le x \le \sqrt{5}$, only $\dfrac{\sqrt{21} - 1}{2}$ is a valid root.

Problem 3.15 Solve for x: $(x - \sqrt{3})x(x + 1) + 3 - \sqrt{3} = 0$.

Answer

$\sqrt{3} - 1$ and $\pm \sqrt[4]{3}$

Solution

This is a cubic equation, and it is difficult to solve directly. So we let $y = \sqrt{3}$. Then

$$(x - y)x(x + 1) + y^2 - y = 0.$$

Expand and factor the left hand side:

$$x^3 - x^2y + x^2 - xy + y^2 - y = (x - y + 1)(x^2 - y) = 0,$$

so we have $x = y - 1$ and $x^2 = y$. Thus the solutions are: $x = \sqrt{3} - 1$ and $x = \pm \sqrt[4]{3}$.

4 Solutions to Chapter 4 Examples

Problem 4.1 Compare: find out which is bigger in each pair, without using a calculator.

(a) $\sqrt{3}$ and $2^{0.8}$

Answer

$\sqrt{3} < 2^{0.8}$.

Solution

Raise both to the 10th power, we have $3^5 = 243 < 256 = 2^8$.

(b) 16^{18} and 18^{16}

Answer

$16^{18} > 18^{16}$

Solution

$16^{18} = 2^{72}$, and $18^{16} = 2^{16} \cdot 3^{32}$, thus we compare 2^{56} and 3^{32}. Now,

$$2^{56} = (2^7)^8 = 128^8 > 81^8 = (3^4)^8 = 3^{32},$$

thus $16^{18} > 18^{16}$.

(c) 2 and $\sqrt{2 + \sqrt{2 + \sqrt{2 + \sqrt{3}}}}$

Answer

$2 > \sqrt{2 + \sqrt{2 + \sqrt{2 + \sqrt{3}}}}$

Solution

Enlarge the right hand side, $\sqrt{2 + \sqrt{2 + \sqrt{2 + \sqrt{3}}}} < \sqrt{2 + \sqrt{2 + \sqrt{2 + \sqrt{4}}}} = 2$.

Problem 4.2 Solve the inequality: $|-x^2 + 2x - 3| < |3x - 1|$

Answer

$(1,4)$

Solution

Since $x^2 - 2x + 3 = (x-1)^2 + 2 > 0$ for all real numbers x, the left hand side

$$|-x^2 + 2x - 3| = |x^2 - 2x + 3| = x^2 - 2x + 3.$$

We solve the inequality in two cases: $x \geq \dfrac{1}{3}$ and $x < \dfrac{1}{3}$.

Case (1): $x \geq \dfrac{1}{3}$. Then

$$\begin{aligned} x^2 - 2x + 3 &< 3x - 1, \\ x^2 - 5x + 4 &< 0, \\ (x-1)(x-4) &< 0. \end{aligned}$$

This gives the solution $1 < x < 4$, which is within the interval $x \geq \dfrac{1}{3}$.

Case (2): $x < \dfrac{1}{3}$. Then

$$\begin{aligned} x^2 - 2x + 3 &< 1 - 3x, \\ x^2 + x + 2 &< 0, \\ \left(x + \frac{1}{2}\right)^2 + \frac{7}{4} &< 0. \end{aligned}$$

This has no solutions.
Thus the final solution is $(1,4)$.

Problem 4.3 Solve the inequality: $|\log_{\sqrt{2}} x - 2| - |\log_2 x - 2| < 2$

Answer

$(1/4, 4)$

Solution

Let $y = \log_2 x$, then $x = 2^y = (\sqrt{2})^{2y}$, so the inequality becomes

$$|2y - 2| - |y - 2| < 2.$$

There are 3 cases: $y \leq 1$, $1 < y \leq 2$, and $y > 2$.
Case (1): $y \leq 1$, then

$$2 - 2y - (2 - y) < 2,$$

so $y > -2$, and the solution for this case is $-2 < y \le 1$.
Case (2): $1 < y \le 2$, then
$$2y - 2 - (2 - y) < 2,$$
so $3y - 4 < 2$, and $y < 2$, thus the solution is $1 < y < 2$.
Case (3): $y > 2$, then
$$2y - 2 - (y - 2) < 2,$$
so $y < 2$, a contradiction, so there is no solution in this case.
Combining Cases (1) and (2), $-2 < y < 2$, so the final solution is $\frac{1}{4} < x < 4$.

Problem 4.4 Solve the inequality: $\dfrac{4x^2}{(1 - \sqrt{1 + 2x})^2} < 2x + 9$

Answer

$$\left[-\frac{1}{2}, 0\right) \cup \left(0, \frac{45}{8}\right)$$

Solution

First of all, $1 + 2x \ge 0$, and $1 - \sqrt{1 + 2x} \ne 0$. Thus, $x \ge -\frac{1}{2}$, and $x \ne 0$.
Let $y = \sqrt{1 + 2x}$, so $y \ge 0$, $y \ne 1$, and $4x^2 = (y^2 - 1)^2$, $2x + 9 = y^2 + 8$, so the inequality becomes
$$\frac{(y^2 - 1)^2}{(1 - y)^2} < y^2 + 8.$$
We know that $y \ne 1$, so
$$\frac{(y^2 - 1)^2}{(1 - y)^2} = \frac{(y + 1)^2 (y - 1)^2}{(y - 1)^2} = (y + 1)^2,$$

then

$$(y+1)^2 < y^2 + 8,$$

$$y^2 + 2y + 1 < y^2 + 8,$$

$$2y < 7,$$

$$y < \frac{7}{2},$$

$$\sqrt{1+2x} < \frac{7}{2},$$

$$1 + 2x < \frac{49}{4},$$

$$x < \frac{45}{8}.$$

Combining with $x \geq -\dfrac{1}{2}$ and $x \neq 0$, the final answer is

$$x \in \left[-\frac{1}{2}, 0\right) \cup \left(0, \frac{45}{8}\right).$$

Problem 4.5 Solve for x: $\dfrac{2}{3}(27^x - 12^x) > 18^x - 8^x$.

Answer

$(-\infty, 0) \cup (1, +\infty)$

Solution

Divide both sides by 8^x,

$$\frac{2}{3}\left(\left(\frac{27}{8}\right)^x - \left(\frac{3}{2}\right)^x\right) > \left(\frac{9}{4}\right)^x - 1.$$

Let $y = \left(\dfrac{3}{2}\right)^x$, and simplify,

$$\begin{aligned}
2(y^3 - y) &> 3(y^2 - 1), \\
2y(y^2 - 1) - 3(y^2 - 1) &> 0, \\
(2y - 3)(y - 1)(y + 1) &> 0.
\end{aligned}$$

Since $y = \left(\dfrac{3}{2}\right)^x > 0$, we have $y + 1 > 0$, so we only need to consider the signs of $2y - 3$ and $y - 1$, therefore the solution for y is $y < 1$ or $y > \dfrac{3}{2}$. Consequently, $x < 0$ or $x > 1$.

Problem 4.6 $\sqrt{x^2 + 4x} \leq 4 - \sqrt{16 - x^2}$

Answer

$\{-4, 0\}$

Solution

We need $16 - x^2 \geq 0$, so $-4 \leq x \leq 4$; in addition, $x^2 + 4x \geq 0$, so $x \leq -4$ or $x \geq 0$. Thus $0 \leq x \leq 4$ or $x = -4$.
It is easy to check that $x = -4$ is a solution.
Now consider the case $0 \leq x \leq 4$:

$$\sqrt{x^2 + 4x} + \sqrt{16 - x^2} \leq 4,$$

squaring both sides (since both sides are nonnegative, squaring both sides does not change the solution set),

$$x^2 + 4x + 16 - x^2 + 2\sqrt{x^2 + 4x} \cdot \sqrt{16 - x^2} \leq 16,$$

which simplified to

$$4x + 2\sqrt{x^2 + 4x} \cdot \sqrt{16 - x^2} \leq 0.$$

Since both terms on the left hand side are nonnegative, they have to be both 0, therefore

$$x = 0.$$

Thus, only $x = 0$ and $x = -4$ are solutions.

Problem 4.7 If x, y, z are positive and $x + y + z = 1$, find the minimum value of

$$\frac{1}{x} + \frac{4}{y} + \frac{9}{z}.$$

Answer

36

Solution

This is a direct application of the Cauchy-Schwarz Inequality.

$$
\begin{aligned}
\frac{1}{x}+\frac{4}{y}+\frac{9}{z} &= \left(\frac{1}{x}+\frac{4}{y}+\frac{9}{z}\right)(x+y+z) \\
&= \left(\left(\sqrt{\frac{1}{x}}\right)^2+\left(\sqrt{\frac{4}{y}}\right)^2+\left(\sqrt{\frac{9}{z}}\right)^2\right)\left((\sqrt{x})^2+(\sqrt{y})^2+(\sqrt{z})^2\right) \\
&\geq (1+2+3)^2 \\
&= 36.
\end{aligned}
$$

Problem 4.8 Assume $a, b, c > 0$, find the minimum values of the following:

$$
\frac{a}{b+c}+\frac{b}{c+a}+\frac{c}{a+b}.
$$

(**Hint:** Add 1 to each fraction, and apply Cauchy-Schwarz Inequality.)

Answer

3/2

Solution

Let

$$
S = \frac{a}{b+c}+\frac{b}{c+a}+\frac{c}{a+b},
$$

add 1 to ach fraction and apply Cauthy-Schwarz In equality:

$$
\begin{aligned}
S+3 &= \frac{a}{b+c}+1+\frac{b}{c+a}+1+\frac{c}{a+b}+1 \\
&= \frac{a+b+c}{b+c}+\frac{a+b+c}{c+a}+\frac{a+b+c}{a+b} \\
&= (a+b+c)\left(\frac{1}{b+c}+\frac{1}{c+a}+\frac{1}{a+b}\right) \\
&= \frac{1}{2}((b+c)+(c+a)+(a+b))\left(\frac{1}{b+c}+\frac{1}{c+a}+\frac{1}{a+b}\right) \\
&\geq \frac{1}{2}(1+1+1)^2 \\
&= \frac{9}{2},
\end{aligned}
$$

so $S \geq \dfrac{3}{2}$.

Problem 4.9 Assume $0 < x < 1$, find the maximum value of $x(1-x^4)$. (Reminder: calculus is not allowed.)

Answer

$\dfrac{4\sqrt[4]{125}}{25}$.

Solution

Let $A = x(1-x^4)$, then

$$
\begin{aligned}
A^4 &= x^4(1-x^4)^4 \\
&= \frac{1}{4}\cdot(4x^4)(1-x^4)^4 \\
&\leq \frac{1}{4}\left(\frac{4x^4+(1-x^4)+(1-x^4)+(1-x^4)+(1-x^4)}{5}\right)^5 \\
&= \frac{1}{4}\left(\frac{4}{5}\right)^5,
\end{aligned}
$$

therefore $A \leq \dfrac{4}{5\sqrt[4]{5}} = \dfrac{4\sqrt[4]{125}}{25}$. Equality occurs when $x = \dfrac{1}{\sqrt[4]{5}}$.

Problem 4.10 Let A, B, C be the 3 angles of $\triangle ABC$. What is the minimum value of $\dfrac{1}{A^2} + \dfrac{1}{B^2} + \dfrac{1}{C^2}$? (Angles are in radians)

Answer

$27/\pi^2$.

Solution

Using AM-GM twice:

$$
\begin{aligned}
\frac{1}{A^2} + \frac{1}{B^2} + \frac{1}{C^2} &\geq 3\sqrt[3]{\frac{1}{A^2 B^2 C^2}} \\
&= \frac{3}{\left(\sqrt[3]{ABC}\right)^2} \\
&\geq \frac{3}{\left(\dfrac{A+B+C}{3}\right)^2} \\
&= \frac{3}{\left(\dfrac{\pi}{3}\right)^2} \\
&= \frac{27}{\pi^2}.
\end{aligned}
$$

Problem 4.11 Find the smallest positive value k such that

$$
\log_{10}(xy) \leq \log_{10} k \cdot \sqrt{\log_{10}^2 x + \log_{10}^2 y}
$$

is true for all $x > 1$ and $y > 1$.

Answer

$10^{\sqrt{2}}$

Solution

Let $a = \log_{10} x$ and $b = \log_{10} y$, and $N = \log_{10} k$, we want to find the minimum value of N such that

$$
a + b \leq N\sqrt{a^2 + b^2}
$$

holds for all $a > 0$ and $b > 0$. By Cauchy-Schwartz Inequality,

$$(a+b)^2 = (1 \cdot a + 1 \cdot b)^2 \le (1^2 + 1^2)(a^2 + b^2) = 2(a^2 + b^2)$$

for all positive a and b, and equality occurs when $a = b$. This means if $N < \sqrt{2}$, the inequality $a + b \le N\sqrt{a^2 + b^2}$ does not hold if $a = b$. Therefore $N \ge \sqrt{2}$, and then $k \ge 10\sqrt{2}$.

Problem 4.12 Find the maximum and minimum of $\sin x + \cos x$.

Answer

$$-\sqrt{2} \le \sin x + \cos x \le \sqrt{2}$$

Solution 1

Rewrite $\sin x + \cos x$ as one sine function,

$$
\begin{aligned}
\sin x + \cos x &= \sqrt{2}\left(\frac{1}{\sqrt{2}}\sin x + \frac{1}{\sqrt{2}}\cos x\right) \\
&= \sqrt{2}(\sin x \cos 45° + \cos x \sin 45°) \\
&= \sqrt{2}\sin(x + 45°).
\end{aligned}
$$

Therefore the minimum value is $-\sqrt{2}$ and the maximum value is $\sqrt{2}$.

Solution 2

Square the expression $\sin x + \cos x$ and use the double-angle formula,

$$(\sin x + \cos x)^2 = \sin^2 x + 2\sin x \cos x + \cos^2 x = 1 + \sin 2x \le 2,$$

thus $|\sin x + \cos x| \le \sqrt{2}$, and so

$$-\sqrt{2} \le \sin x + \cos x \le \sqrt{2}.$$

Solution 3

Apply the Cauchy-Schwarz Inequality on a more general expression $a\sin x + b\cos x$, using the fact that $\sin^2 x + \cos^2 x = 1$,

$$(a\sin x + b\cos x)^2 \le (a^2 + b^2)(\sin^2 x + \cos^2 x) = a^2 + b^2,$$

thus $|a\sin x + b\cos x| \le \sqrt{a^2 + b^2}$, and so

$$-\sqrt{a^2 + b^2} \le a\sin x + b\cos x \le \sqrt{a^2 + b^2}.$$

For this problem, $a = b = 1$, therefore

$$-\sqrt{2} \le \sin x + \cos x \le \sqrt{2}.$$

Problem 4.13 Find the range of the function $y = \sqrt{4x - 1} + \sqrt{2 - x}$.

Answer

$$\left[\frac{\sqrt{7}}{2}, \frac{\sqrt{35}}{2}\right]$$

Solution

The domain for x is $\frac{1}{4} \le x \le 2$.

Make a change of variables, $t = x - \frac{1}{4}$, then

$$y = 2\sqrt{t} + \sqrt{\frac{7}{4} - t},$$

where $0 \le t \le \frac{7}{4}$.

Now let $t = \frac{7}{4}\sin^2\alpha$ where $0 \le \alpha \le \frac{\pi}{2}$, then $\frac{7}{4} - t = \frac{7}{4}\cos^2\alpha$, and (after simplification)

$$y = \sqrt{7}\sin\alpha + \frac{\sqrt{7}}{2}\cos\alpha.$$

Now we write the expression as one single sine function. Since

$$\left(\sqrt{7}\right)^2 + \left(\frac{\sqrt{7}}{2}\right)^2 = 7 + \frac{7}{4} = \frac{35}{4},$$

let θ be the angle in $\left[0, \frac{\pi}{2}\right]$ such that $\sin\theta = \dfrac{\sqrt{7}/2}{\sqrt{35/4}} = \dfrac{\sqrt{5}}{5}$, (in other words, $\theta = \arcsin\dfrac{\sqrt{5}}{5}$), then

$$y = \frac{\sqrt{35}}{2}\sin(\alpha + \theta),$$

where $\theta \le \alpha + \theta \le \dfrac{\pi}{2} + \theta$.

Therefore, the maximum value is reached when $\alpha + \theta = \dfrac{\pi}{2}$ and $y = \dfrac{\sqrt{35}}{2}$, and the minimum value is reached at one of the end points of the interval: $\alpha + \theta = \theta = \arcsin\dfrac{\sqrt{5}}{5}$ and $y = \dfrac{\sqrt{7}}{2}$. (At the other end point, $y = \sqrt{7}$, which is larger.)

Problem 4.14 Solve for x: $\dfrac{x}{\sqrt{1+x^2}} + \dfrac{1-x^2}{1+x^2} > 0$.

Answer

$x > -\sqrt{3}/3$.

Solution

Let $x = \tan\theta$, $-\dfrac{\pi}{2} < \theta < \dfrac{\pi}{2}$, then

$$
\begin{aligned}
\frac{x}{\sqrt{1+x^2}} + \frac{1-x^2}{1+x^2} &= \frac{\tan\theta}{\sqrt{1+\tan^2\theta}} + \frac{1-\tan^2\theta}{1+\tan^2\theta} \\[2mm]
&= \frac{\tan\theta}{\sec\theta} + \frac{1-\tan^2\theta}{\sec^2\theta} \\[2mm]
&= \sin\theta + \cos^2\theta - \sin^2\theta \\[2mm]
&= \sin\theta + 1 - 2\sin^2\theta \\[2mm]
&= -(2\sin\theta + 1)(\sin\theta - 1) \\[2mm]
&> 0.
\end{aligned}
$$

Since $-\dfrac{\pi}{2} < \theta < \dfrac{\pi}{2}$, $\sin\theta - 1 < 0$, therefore $2\sin\theta + 1 > 0$, which means $\sin\theta > -\dfrac{1}{2}$, so $-\dfrac{\pi}{6} < \theta < \dfrac{\pi}{2}$, and thus $x > -\dfrac{\sqrt{3}}{3}$.

Problem 4.15 If $2x + y \ge 1$, find the minimum value for $u = y^2 - 2y + x^2 + 4x$.

Answer

$-9/5$

Solution

Rewrite the equation $u = y^2 - 2y + x^2 + 4x$ as

$$(x+2)^2 + (y-1)^2 = u + 5.$$

This means that $\sqrt{u+5}$ is the distance between points (x, y) and $(-2, 1)$. Since the point (x, y) satisfies $2x + y \geq 1$, it falls on the upper right side of the line $2x + y = 1$. On the other hand, the point $(-2, 1)$ falls on the lower left side of the line $2x + y = 1$, so the minimum distance equals the distance from the point $(-2, 1)$ to the line $2x + y = 1$. The formula of the distance from a point (x_0, y_0) to a line $Ax + By + C = 0$ is (verify it!)

$$d = \frac{Ax_0 + By_0 + C}{\sqrt{A^2 + B^2}}.$$

In this question, the point is $(-2, 1)$ and the line is $2x + y - 1 = 0$, so the minimized distance is

$$\sqrt{u+5} = \frac{2(-2) + 1 - 1}{\sqrt{2^2 + 1^2}} = -\frac{4}{\sqrt{5}}.$$

thus

$$u + 5 = \left(-\frac{4}{\sqrt{5}}\right)^2 = \frac{16}{5},$$

so

$$u = \frac{16}{5} - 5 = -\frac{9}{5}.$$

5   Solutions to Chapter 5 Examples

Problem 5.1 Find one solution for $x^{x^2} = 2$ by observation.

Answer

$\sqrt{2}$ or $-\sqrt{2}$

Solution

It is not hard to see that if $x^2 = 2$, the equation holds. Therefore $x = \sqrt{2}$ is one solution. Also $x = -\sqrt{2}$ is another solution.

Problem 5.2 Find the sum of the roots of $x^2 - 2000|x| = 2000$.

Answer

0

Solution

If x_0 is a root, then $-x_0$ is also a root. So the sum is 0.

Problem 5.3 Solve for x, using the method of Completing the Square.

$$x + \frac{x}{\sqrt{x^2 - 1}} = \frac{35}{12}.$$

Hint: first square both sides.

Answer

$\dfrac{5}{3}$ and $\dfrac{5}{4}$

Solution

Clearly $x > 1$. Square both sides,

$$x^2 + \frac{2x}{\sqrt{x^2 - 1}} + \frac{x^2}{x^2 - 1} = \frac{1225}{144},$$

which is

$$\frac{x^4}{x^2 - 1} + \frac{2x^2}{\sqrt{x^2 - 1}} = \frac{1225}{144}.$$

Add 1 to both sides to complete the square,

$$\frac{x^4}{x^2-1} + \frac{2x^2}{\sqrt{x^2-1}} + 1 = \frac{1225}{144} + 1,$$

which is

$$\left(\frac{x^2}{\sqrt{x^2-1}} + 1\right)^2 = \frac{1369}{144}.$$

Since $x > 1$, take the positive square root,

$$\frac{x^2}{\sqrt{x^2-1}} + 1 = \frac{37}{12}.$$

Thus

$$\frac{x^2}{\sqrt{x^2-1}} = \frac{25}{12}.$$

Let $y = \sqrt{x^2-1}$, then the equation becomes

$$\frac{y^2+1}{y} = \frac{25}{12}.$$

Solve the quadratic equation in y, $y = \frac{4}{3}$ or $\frac{3}{4}$. Thus

$$\sqrt{x^2-1} = \frac{4}{3} \text{ or } \frac{3}{4}.$$

Solve for x, taking positive square roots, $x = \frac{5}{3}$ or $\frac{5}{4}$. Checking with the original equation, these are both roots.

Note: Another common technique is trig substitution: $x = \dfrac{1}{\sin\theta} \,(0° < \theta < 90°)$.

Problem 5.4 Solve for x: $\sqrt{5-x} + \sqrt{2+x} = \sqrt{5} + \sqrt{2}$.

Answer

0 or 3

Solution

Squaring,

$$\sqrt{5-x} \cdot \sqrt{2+x} = \sqrt{10}.$$

Using Vieta's Theorem, $\sqrt{5-x}$ and $\sqrt{2+x}$ are roots of quadratic equation

$$t^2 - (\sqrt{5} + \sqrt{2})t + \sqrt{10}.$$

Solve to get two solutions: $x = 0$ and $x = 3$.

Problem 5.5 Solve for x:

$$(6x+7)^2(3x+4)(x+1) = 6.$$

Hint: Convert $3x+4$ to $6x+8$ and $x+1$ to $6x+6$.

Answer

$-\dfrac{2}{3}$ and $-\dfrac{5}{3}$

Solution

The equation is equivalent to

$$(6x+7)^2(6x+8)(6x+6) = 72.$$

Let $y = 6x+7$, then

$$y^2(y+1)(y-1) = 72,$$

so

$$y^4 - y^2 - 72 = 0.$$

Solve to get $y^2 = 9$ (throw away -8), and then $6x+7 = \pm3$, thus $x = -\dfrac{2}{3}$ or $x = -\dfrac{5}{3}$.

Problem 5.6 Solve for x:

$$(3x^2 - 2x + 1)(3x^2 - 2x - 7) + 12 = 0.$$

Answer

$1, -\dfrac{1}{3}, -1, \dfrac{5}{3}$

Solution

Let $y = 3x^2 - 2x - 3$, then

$$(y+4)(y-4) + 12 = 0,$$

which simplifies to $y^2 - 4 = 0$, thus $y = \pm 2$. Therefore

$$3x^2 - 2x - 3 = -2$$

or

$$3x^2 - 2x - 3 = 2.$$

Solving for x, then the four roots are $1, -\dfrac{1}{3}, -1, \dfrac{5}{3}$.

Problem 5.7 Find real solutions:

$$\sqrt{x-1} + 2\sqrt{y-4} + 3\sqrt{z-9} + 4\sqrt{w-16} = \frac{1}{2}(x+y+z+w).$$

Answer

$x = 2, y = 8, z = 18, w = 32$

Solution

Completing the squares,

$$(\sqrt{x-1} - 1)^2 + (\sqrt{y-4} - 2)^2 + (\sqrt{z-9} - 3)^2 + (\sqrt{w-16} - 4)^2 = 0,$$

thus each square has to be 0. So $x = 2, y = 8, z = 18, w = 32$.

Problem 5.8 Find all real roots for $x^2 - 2x\sin\dfrac{\pi x}{2} + 1 = 0$.

Answer

± 1

Solution

To find the real roots, we calculate the discriminant

$$4\sin^2\frac{\pi x}{2} - 4 \geq 0.$$

Therefore

$$\sin^2 \frac{\pi x}{2} \geq 1.$$

This requires that

$$\sin^2 \frac{\pi x}{2} = 1.$$

So the equation becomes

$$x^2 \pm 2x + 1 = 0,$$

There are two solutions: $x = \pm 1$.

Problem 5.9 Given $a > b$, solve for x:

$$\sqrt{a-x} + \sqrt{x-b} = \sqrt{a-b}.$$

Answer

$x = a$ or $x = b$

Solution

We use a geometric interpretation.

It is required that $b \leq x \leq a$, and it is easy to see that $x = a$ and $x = b$ are both roots.

If $b < x < a$, consider $\sqrt{a-x}$ and $\sqrt{x-b}$ as the two legs of a right triangle, then $\sqrt{a-b}$ is the hypotenuse. By Triangle Inequality, $\sqrt{a-x} + \sqrt{x-b} > \sqrt{a-b}$ for all $b < x < a$. Therefore the only two roots are $x = a$ and $x = b$.

Problem 5.10 Let real numbers $x_1, x_2, \ldots, x_n$ satisfy

$$\frac{x_1}{x_1^2 + 1} = \cdots = \frac{x_n}{x_n^2 + 1},$$

$$x_1 + \cdots + x_n + \frac{1}{x_1} + \cdots + \frac{1}{x_n} = \frac{10}{3}.$$

Find the value of x_n.

Answer

3 or 1/3

Solution

Take reciprocals in the first equation,

$$x_1 + \frac{1}{x_1} = \cdots = x_n + \frac{1}{x_n},$$

thus from the second equation, $n\left(x_1 + \frac{1}{x_1}\right) = \frac{10}{3}$, that is,

$$nx_1^2 - \frac{10}{3}x_1 + n = 0.$$

Since x_1 is a real number, the discriminant $\left(-\frac{10}{3}\right)^2 - 4n^2 \geq 0$, so $n \leq \frac{5}{3}$. But n is a positive integer, thus $n = 1$. Finally, $x_1 = 3$ or $\frac{1}{3}$.

Problem 5.11 Solve for x, y:

$$\begin{cases} \sqrt{3-y} = \sqrt{x} + \sqrt{x-y}, \\ \sqrt{1-y} = \sqrt{x} - \sqrt{x-y}. \end{cases}$$

Answer

$x = 1, y = \dfrac{3}{4}$

Solution

Multiplying the two equations, $\sqrt{(3-y)(1-y)} = y$, then $y = \dfrac{3}{4}$, and then $x = 1$.

Problem 5.12 (IMO 1959) Find all possible values of x satisfying the equation

$$\sqrt{x + \sqrt{2x-1}} + \sqrt{x - \sqrt{2x-1}} = \sqrt{2}.$$

Answer

$\dfrac{1}{2} \leq x \leq 1$

Solution

Let $y = \sqrt{2x-1}$. Then $x = \dfrac{y^2+1}{2}$. Thus

$$\sqrt{\frac{y^2+2y+1}{2}} + \sqrt{\frac{y^2-2y+1}{2}} = \sqrt{2},$$

thus

$$y+1+|y-1| = 2.$$

If $y \geq 1$, $2y = 2$, so $y = 1$, thus $x = 1$.

If $y < 1$, $y+1+1-y = 2$, always true. Also note that $y = \sqrt{2x-1} \geq 0$, so $0 \leq y < 1$, which means $\dfrac{1}{2} \leq x < 1$.

Combining the two cases, $\dfrac{1}{2} \leq x \leq 1$.

Problem 5.13 Find real solutions to the system

$$\begin{cases} y = \sqrt{x - \dfrac{1}{x}} + \sqrt{1 - \dfrac{1}{x}}, \\ x = \sqrt{y - \dfrac{1}{y}} + \sqrt{1 - \dfrac{1}{y}}. \end{cases}$$

Hint: From the first equation, calculate $2x - 2y$, and complete the squares. Then obtain a similar result from the second equation.

Answer

$$x = y = \frac{1+\sqrt{5}}{2}$$

Solution

From the first equation, calculate $2x - 2y$, and complete the squares,

$$2x - 2y = 2x - 2\sqrt{x-\frac{1}{x}} - 2\sqrt{1-\frac{1}{x}} = \left(\sqrt{x-\frac{1}{x}}-1\right)^2 + \left(\sqrt{x-1}-\frac{1}{\sqrt{x}}\right)^2 \geq 0,$$

thus $x \geq y$. Similarly, $y \geq x$. So $x = y$, and then

$$\left(\sqrt{x-\frac{1}{x}}-1\right)^2 + \left(\sqrt{x-1}-\frac{1}{\sqrt{x}}\right)^2 = 0,$$

which means

$$\sqrt{x - \frac{1}{x} - 1} = \sqrt{x - 1} - \frac{1}{\sqrt{x}} = 0,$$

solve and we get $x = \dfrac{1 + \sqrt{5}}{2}$, and thus

$$x = y = \frac{1 + \sqrt{5}}{2}.$$

Problem 5.14 Find **positive** solutions:

$$\begin{cases} x^2 + y^2 + xy = 1, \\ y^2 + z^2 + yz = 3, \\ z^2 + x^2 + zx = 4. \end{cases}$$

Hint: Is there any geometric interpretation?

Answer

$$(x, y, z) = \left(\frac{2}{\sqrt{7}}, \frac{1}{\sqrt{7}}, \frac{4}{\sqrt{7}} \right)$$

Solution

First consider a triangle ABC. Let P be the point in $\triangle ABC$ such that

$$\angle APB = \angle BPC = \angle CPA = 120° \text{(the Fermat point)}.$$

Let $AP = x$, $BP = y$, and $CP = z$, then

$$\begin{array}{ccccc} AB^2 & = & x^2 + y^2 + xy & = & 1, \\ BC^2 & = & y^2 + z^2 + yz & = & 3, \\ CA^2 & = & z^2 + x^2 + zx & = & 4, \end{array}$$

therefore $AB = 1$, $BC = \sqrt{3}$, and $CA = 2$, and this triangle is a 30-60-90 triangle. It is easy to find the area of the triangle: $\dfrac{1}{2} \cdot \sqrt{3} \cdot 1 = \dfrac{\sqrt{3}}{2}$. Also the area equals

$$\frac{1}{2} xy \sin 120° + \frac{1}{2} yz \sin 120° + \frac{1}{2} zx \sin 120°,$$

so
$$\frac{\sqrt{3}}{4}(xy + yz + zx) = \frac{\sqrt{3}}{2},$$

thus
$$xy + yz + zx = 2.$$

Add the three equations,
$$2(x^2 + y^2 + z^2) + (xy + yz + zx) = 8,$$

and then
$$2(x+y+z)^2 = 2(x^2 + y^2 + z^2) + 4(xy + yz + zx) = 14,$$

hence
$$x + y + z = \sqrt{7}.$$

Now subtract the first equation from the second,
$$z^2 - x^2 + y(z - x) = 2,$$

which is
$$(x + y + z)(z - x) = 2.$$

Similarly, subtract the first equation from the third,
$$z^2 - y^2 + x(z - y) = 3,$$

which is
$$(x + y + z)(z - y) = 3.$$

So we get
$$z - x = \frac{2}{\sqrt{7}}, \quad z - y = \frac{3}{\sqrt{7}}.$$

Solving for x, y, z, then
$$(x, y, z) = \left(\frac{2}{\sqrt{7}}, \frac{1}{\sqrt{7}}, \frac{4}{\sqrt{7}} \right).$$

Problem 5.15 Find all **integer roots** of $\dfrac{x+y}{x^2 - xy + y^2} = \dfrac{3}{7}$.

Answer

$(4,5), (5,4)$

Solution

The equation becomes

$$7(x+y) = 3(x^2 - xy + y^2),$$

so

$$3x^2 - 3xy + 3y^2 - 7x - 7y = 0.$$

Treat it as a quadratic equation in x,

$$3x^2 - (3y+7)x + (3y^2 - 7y) = 0.$$

For the equation to have real solutions, the discriminant should be nonnegative:

$$(3y+7)^2 - 4 \cdot 3 \cdot (3y^2 - 7y) \geq 0,$$

so

$$9y^2 + 42y + 49 - 36y^2 + 84y \geq 0,$$

which is

$$27y^2 - 126y - 49 \leq 0.$$

The solution to this quadratic inequality is

$$\frac{21 - 14\sqrt{3}}{9} \leq y \leq \frac{21 + 14\sqrt{3}}{9}$$

The integer values in this interval are $0, 1, 2, 3, 4, 5$. Only $y = 4$ nd $y = 5$ give integer solutions for x, and solve to get $(x, y) = (4, 5)$ or $(5, 4)$.

Made in the USA
Monee, IL
02 July 2022